THE
ASTRO-CAFÉ
CONSPIRACY

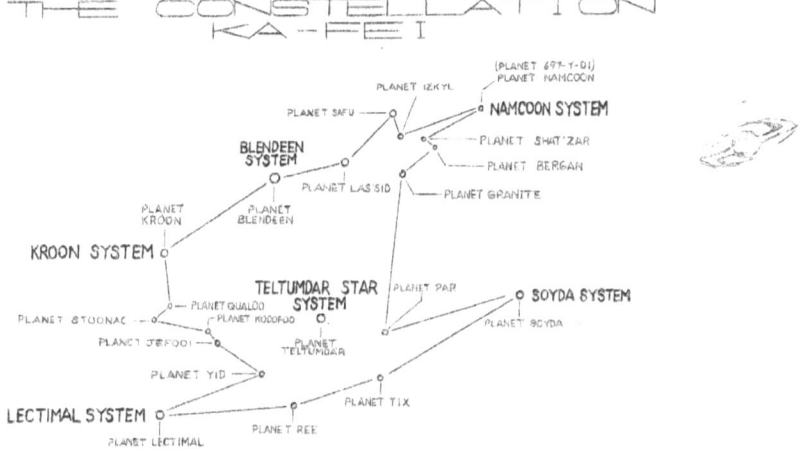

THE
ASTRO-CAFÉ
CONSPIRACY

Brad W. Ellis and Victor M. Silva
Edited By: Ian McGhie
Cover Illustration: Brad W. Ellis
Graphic Design: Vast

Library of Congress Control Number:		2011914920
ISBN:	Hardcover	978-1-4653-5567-6
	Softcover	978-1-4653-5566-9
	Ebook	978-1-4653-5568-3

Contact Details:

AstroCaféDomain@facebook.com
follow us on twitter @AstroCaféDomain

To order additional copies of this book, contact:
Xlibris Corporation
1-888-795-4274
www.Xlibris.com
Orders@Xlibris.com
103242

CONTENTS

Synopsis

In the year 2032, a ship of immense penetrated the earths' atmosphere and (due to the damage bestowed upon it by the United Earth Forces Stellar Defense Grid for potential planet threatening asteroids) crashed-landed thirty miles north of Ulan Bator, Mongolia, creating a fireball large enough to be seen from Russian boarder line. Only two beings in a fifty-mile radius survived the impact, both were passengers.

One was *Galyymann* from the planet Granitan, a renegade priest of Teltumdar sect, a religion unknowingly shared and practice by man and Granites alike for hundreds if not thousands of years. The other was his cargo, a six foot tall living coffee bean possessing extraordinary powers, one of the twelve Beans of Blendeen who calls itself *Ononot.*

There ejection pod was found by a poor fisherman's daughter named *Omi* off the coast of Okinawa, Japan several days later who befriended the priest. Galyymann however kept Ononot and his alien identity a secret fearing it would frighten the beautiful Omi away.

CHAPTER 1

Ground Zero

The time is about four o' clock in the morning in Okinawa, Japan and Mr. Akira Yoshiuchi of the Yoshiuchi Factory gives the order over the radio start the engines of his three remaining fishing travelers; The Deep Neptune, The Coral and The Rising Sun to head out for another days worth of fishing.

Akira: The weather look's okay, but I don't trust it . . .

Omi: Why Papa?

Akira: Because no matter what the sky may look like, there's always something ready to fall out of it, that's why!

Omi: Everything that falls from the sky isn't always bad . . . What about the rain? If it didn't rain we'd all be in trouble.

Akira: Well . . . maybe so, but I still don't trust the weather. It doesn't care what you have planned for the day; if it decides to ruin it . . . look out!

Omi: The water's still . . . maybe we're the first ones out today.

Akira: I hope so 'cause if we are, I'm heading to my secret spot. I can't take any chances of other ships tailing me there.

Omi: Oh daddy . . . everyone in the fishing industry knows about your silly secret spot and nobody cares. So what are you worried about?

Akira: Sweetheart . . . I've been fishing these waters since I was a kid and I'm the only one who knows where the secret spot is. You mean to tell me that after all these years you haven't figured out your father yet?

Omi: What are you talking about?

Akira: I guess you really wouldn't know. O.K. young lady, papa's gonna let you in on his secret.

My father created that secret spot many, many years ago by, baiting this area with all kinds of undersea goodies for the fish to come and eat. The reason that it was kept such a secret is because the baiting takes place deep underwater in specially made baiting cages.

Just before we set out in the morning, I take a small boat, put on a scuba gear and bate the cages. This brings the fish running to the secret spot every time. I think maybe since this has been going on for so many years that most of the fish around here expect the bait cages to be ready at certain times.

Anyway, I head back in and prepare to get the fleet going . . . or what's left of it, that's why certain days this old man is up and dressed before you.

Omi: What a smarty! I would have never guesses!

Akira: I know. The larger industries wonder why I'm still in business. They need to understand that it takes more than fancy pants equipment to fish.

Omi: You hiding anymore secrets from me?

Akira: Yeah, I also like to wear woman's underwear, it's so comfortable and . . .

The foghorn sounds and both Omi and her father head up to the upper deck to take a look-see.

Akira: See what I mean? Where in the Hell did all the fog come from? A few minutes ago we . . .

Seaman1: Captain, we're almost at the designated coordinates.

Akira: Good, give the order to drop the nets to the required depth and standby.

As the seamen go to carry out captain Akira's orders, the water starts to get a little choppy. A few moments later the collision bell rings.

Seamen2: Captain . . . ! There's a huge ship heading our way and it's not altering its course!!

Seamen3: Everyone!! Hold on to something and brace for impact!!

Akira: Hard to port yeoman, hard to port!!

Akira races up to the bridge, pushes the sailor aside and whips the wheel to port side of the boats, simultaneously throwing the engines in reverse. The Deep Neptune scrapes up against the sides of the huge passing boat and almost capsizes. Then their nets get tangled in something connected to the passing vessel and are being pulled apart.

> Seamen1: Captain, our nets are tangled in the other vessels and they're dragging us!!
> Omi: That thing is a fishing boat! It's gigantic!!
> Akira: Omi . . . get down below right now and see if there's any damage . . . you go and cut us free!! If you don't we'll go under for sure!! You . . . turn us about full speed and follow that boat!!
> Seamen2: Aye, aye sir!!

Omi runs below and checks each cabin for structural damage, but finds none. Meanwhile, above deck the crew of The Deep Neptune struggles desperately to cut their nets free before they are pulled under. Akira sends out a distress call just in case there is irreparable damage and then contacts the remainder of his fleet to see if they had suffered the same fate.

At approximately zero-four hundred hours and forty-nine minutes eastern standard time, an unidentified object traveling at twenty-eight thousand miles per minute appear out of nowhere on The United Earth Forces Stellar Defense System shortly after on NORAD's early warning missiles defense grid.

> MIL. Officer 1: What the Hell is that thing?!
> MIL. Officer2: Whatever it is . . . it's big, fast and hot!!
> MIL. Officer1: Contact NORAD and find out if they're getting the same thing . . . if so then switch the Stellar Defense on . . .
> MIL. Officer2: Sir, yes sir!

The call made by the officer confirmed that there was a massive unknown celestial body on a conclusion course with earth. Necessary action had to be taken immediately.

> MIL. Officer3: Stellar Defense System activated sir . . . prepare the Washington for the fireworks display.

The system was activated and began firing it's large particle beam cannons and launching its missile, but to no avail. The object maneuvered out of the way of almost every shot fired.

> MIL. Officer1: Impossible!! What is that thing?
> MIL. Officer3: It moved by itself . . . like it's being piloted!!

MIL. Officer2: It can't be a space craft, look at it!

MIL. Officer1: Contact the Pentagon. We've got to get an interception squadron up there and engage this thing before it hits us! Also contact NORAD and tell them to try to bring that thing down in pieces!!

MIL. Officer2: Done sir!!

MIL. Officer3: We've got and B.T.A. on that bogie sir, twenty-two minutes.

MIL. Officer1: Great . . . that gives us just enough time to do squat!!

Back on board The Deep Neptune, things were starting to calm down. The nets were severed just in time and no serious damage was done to the smaller trawler.

Seamen4: That's it, I quit!!

Seamen2: Don't talk like that, accidents do happen you know.

Seamen4: You call what just happened an accident?! Are you crazy? That bastard tried to sink us!!

Seamen3: Stop being so damn paranoid! Last week, that bastard tried to ram us and send us to the bottom of the ocean! That's it I tell you! Turn this crate around and head back to shore 'cause I'm outta here!

Omi: What's going on here?

Seamen2: Mamuro wants to quit, this time I think he means it.

Omi: Oh Mamuro's such a baby, one little accident and he falls apart like a house of cards or something.

Seamen2: You have to admit that was a little strange. I mean they didn't even slow down, not even after the collision.

Omi: Their boat was so big they didn't even see us. Besides, we're O.K., no one was hurt.

Seamen4: Yeah, this time. Next time they might not miss!!

Omi: Are you trying to give my father bad luck!

Seamen4: Your father already has bad luck!

Omi: Go to Hell Mamuro!!

Seamen4: No problem I just don't want to go at the same time as your father!! Maybe I should just go and work for the competition! Well, you can't call them that because we are not any competition to them!

Omi: Bastard!

Akira: Stop it, the both of you! Don't worry Mamuro; we're heading back to shore as we speak. That huge fishing ship made off with our catch and our nets. We won't be going out until repairs can be made and I can find out what that was all about.

Omi: Daddy, you don't believe what this idiot is saying? That was intentional.

Akira: Omi, their running lights were off when they rammed us. I just got the word from The Rising Sun that they just turn them on.

Seamen4: See . . . ! I told yah

Omi: Shut up!!! Papa, you're a good man. No one would ever want to bring harm to you. It seems that all you do is help people out all the time. You even keep Mamuro on even though everyone knows how much of a slacker he is. People love and respect you for all the hard work you do. You ran a business while raising up a little girl all by yourself.

Akira: I think I did a fine job too.

Seamen3: Hey everybody . . . look at that!!

Something large and bright streaked across the morning sky followed by a large sonic boom. Everyone looked up in silence at the object as it traveled along over head, dropping closer and closer to Earth. Just then Omi spotted something glint off of the morning sun's rays.

Omi: What's that?

Akira: What are you talking about? Where?

Omi: Right over there . . . falling out of the sky. It looks like it might hit that boat over there.

Seamen2: I think you're right! We've got to warn them over the radio!

But it was too late. The object hit the stem of the ill-fated vessel and it exploded into a ball of flames. The crew of The Deep Neptune scrambled to the side where the explosion could be seen.

Seamen3: Oh my goodness, was that The Coral that just exploded? I couldn't see it!

Seamen5: No way!! The Coral's on the other side of us! That was someone else!

Seamen6: Captain . . . don't you think we should get over there and search for survivors?

Akira: With an explosion like that, I doubt there'd be any. Let's go take a gander anyway. Kenji, radio the harbor patrol and tell them what happened. The rest of you prepare to get underway. Omi, make sure we have what we need for: first aid.

Back at The United Earth Forces Stellar Defense System Headquarters located in the heart of Canada not far from the Artic Circle, the tracking of the object continued.

> MIL. Officer2: They didn't get the fighters up in time, the computer's trying to estimate a target area O.K. it's going to hit somewhere in outer Mongolia!
> MIL. Officer2: A populated area?
> MIL. Officer3: Yes . . . A few scattered villages, but populated none the less.
> MIL. Officer2: Can I have an estimated loss of life count?
> MIL. Officer3: I'm working on it . . .
> MIL. Officer1: Probably a couple of thousand, maybe less.
> MIL. Officer2: Anything we can do to warn them?
> MIL. Officer1: That's a negative . . . lets just bow our heads in silence gentlemen.
> MIL. Officer3: God help them . . .

The explosion made by the object striking the Earth could be seen from space. Everyone within a twenty-mile radius was killed. Meanwhile, Akira Yochiuchi and his fleet of fishing trawlers converged on the spot where they saw the ship explode and searched for survivors.

> Omi: I see some one . . . over here on the starboard side!!
> Seamen3: She's right. I can see him. He's holding on to something. He's slipped under!!
> Omi: Somebody grab him!!

Omi seeing the man drowning dives in and swims towards the object he was clinging to.

> Akira: Omi! What in the Hell do you think you're doing?!
> Seamen1: She's too far out now captain, she can't hear you!
> Seamen3: That water's freezing! If she stays in too long, she'll die!
> Seamen4: She's crazy!
> Akira: Maybe but she's still my daughter and my responsibility! I've got to go after her!
> Seamen4: Wait! You can't! If you die . . . , who's going to pay me this week's salary?
> Seamen3: Mamuro you idiot, shut the Hell up and give me a hand with this extra net, I've got an idea.

Akira: Gonna try and fish 'em out huh? Great idea!!
Seamen3: It's what we do best, right?

The captain and crew of The Deep Neptune move quickly and set up the extra net so they can scoop Omi and the drowning man out of the frigid water, They managed to do so just before Omi goes under with the stranger in her arms. They bring the two, along with some wreckage on board and begin to resuscitate the now unconscious man while her father comforts Omi.

Akirn: Don't you ever do that again!! You had me worried sick!! Are you trying to give this old man a heart attack or something?

Omi: Papa I'm sorry. Everybody was just standing around watching . . . , I . . . , I just couldn't stand it anymore!! I had to jump in . . .

Akira: I'm not really angry with you; just a little scared that's all. You know how protective I am of you. Your mother would never forgive me if something were to happen to you.

Omi: You don't think she would have been proud of me?

Akira: You mean before or after she passed out? Look, your mom would not have expected anything less from you; you are a Yoshiuchi you know. Our ancestor was very smart and brave and was honored by many. I'm so ashamed that I can't seem to keep that going, it drives me crazy!

Omi: Don't be silly! You're three times the man of any man on board any ship!

Akira: A while ago I probably would have agreed with you . . . , but now . . .

Omi: Rest now papa We should be getting back to shore pretty soon and you're going to need your energy because you're going to be interviewed by all sorts of television crews and the Harbor Patrol and Police . . . , just like the hero I know you are.

Akira: Thank you sweet heart, I think this old man could use some rest.

Seaman 1: Captain, the Harbor Patrol's about five minutes off our port side.

Akira: G:ood, get them on the radio and tell them we have a survivor on board that needs medical assistance.

Omi: Papa, get some rest. I'll go and talk to the Harbor Patrol.

Seamen 1: But what if he needs to speak to the captain?

Omi: After I tell them he's recuperating from diving into the frozen waters and saving a man's life they'd be a little more understanding don't you think?

Seamen 1: ·Oh . . . I get it. I'll just go and tell the rest of the crew what we're talking about.

Meanwhile, the whole world has been told of the disaster that happened in Mongolia Also, a worldwide military, science and rescue expedition has been dispatched to the impact area to assess the damage and find out what hit the Earth. Dr. Cocoa and his team of Stellar Defense System technicians have been called into an emergency meeting at The United Nations Headquarters in New York City.

Member 1: Dr. Kaneohe Cocoa, we here at the United Nations are in deep mourning from the past few events that have happened over the years. The continuous fighting in the Gulf, the attacks on Taiwan by Chinese extremists, the constant growth in our world's hunger population . . . , all of these pale in comparison with the most recent event; the near destruction of a country by the asteroid; wouldn't you say?

DR. COCOA: Absolutely sir, I couldn't agree with you more.

Member1: Good. I hope you also realized how much faith the world has . . . , we had in the modifications . . . , your modifications on the United Earth Forces Stellar Defense Grid.

DRCOCOA: Umm, yes sir, I suppose I do . . . , I mean did.

Tech1: Look out doc, I think they are looking for someone to blame.

DR. COCOA: I know John . . . , do you think I'm stupid!?

MEMBER1: Then I hope you understand what this meetings all about then.

DR. COCOA: No sir . . . , I'm afraid I do not.

MEMBER2: Then we will explain. Your incompetence towards the project mentioned caused the lives of thousands of people in Mongolia to be lost, not to mention the people that are going to die from the poisonous cloud released into our atmosphere . . .

TECH2: Here it comes . . .

DR. COCOA: Wait just one minute here!! You can't blame me for that!! I didn't voluntarily allow that thing to fall from the sky; neither did I kill anyone . . .

MEMBER3: We are aware of the fact that the Stellar Defense Grid was supposed to be in top working order approximately three weeks ago!

DR. COCOA: It was! What am I saying it still IS!! There isn't anything wrong with it!! It passed all of my system tests and . . .

MEMBER 2: Yes Dr. Cocoa . . . your system tests. But what about ours?

TECH 2: Your systems tests?! I thought we worked for YOU?!!

MEMBER 1: When we say ours we mean the entire world. You forget this is The United Nations sir, you, ALSO forget that you are speaking out of turn, now please remain seated!

DR. COCOA: Please, you don't understand the blame you're holding over my head . . .

MEMBER 1: And you don't understand the tragedy that has just recently occurred Dr. Cocoa. We know you did not intentionally cause all of that destruction but . . .

DR. COCOA: INTENTIONALLY!?!

Member 1: BUT . . . , we have decided that if The Stellar Defense Grid was in full operational order this tragedy would have nev . . .

Tech 1: The only tragedy here today gentlemen is the condemnation of a man who worked his ass off on a project originally designed to detect and destroy hostile bodies in deep space, not a few thousand miles . . .

Member 2: I am going to ask you to explain to your technical team that screaming out of turn is unacceptable in front of . . .

Dr. Cocoa: The only thing I'll explain to them is that it really doesn't pay to care about your work because in the long run it doesn't matter in the end!!

Tech 3: What a cheap shot!! Go to Hem!!

Member 1: One more outburst like that young man and I'll have you removed!!!

DR. COCOA: Enough!!! It looks like you gentlemen have made up your mind about this so if you are not going to arrest me then I think my co-workers and I are competent enough to remove ourselves thank you!

Member 2: Wait a moment; this meeting is FAR from over!!

DR. Cocoa: Not for me it isn't, good-bye gentlemen. If you need anything you'll know where to find me . . . , at home with my wife. Let's go John, Gregory . . . Wong.

Member 1: Gentlemen, remain in your seats, PLEASE!!! This meeting is NOT over!!!

The next day at the Okinawa General hospital, Omi visits the stranger who still hasn't regained full consciousness.

Omi: Don't worry; I won't leave your side until somebody finds out who you are. You're so handsome. It would be a shame if you didn't wake up . . . , I'm sure there's someone out there looking for you that's worried sick. I wonder why you were the only one that survived . . . , lucky I guess.

Nurse 'Excuse me ma'am . . . , I'm sorry but visiting hours are up.'

OMI 'Okay, I'll be leaving soon. Um . . . , still no word of any relatives or anything like that?

Nurse: I'm afraid not. He isn't even Japanese so I suppose it will be a while before anyone shows up. Anyway, you're doing a pretty good job of taking care of him, why don't I wrap him up for you so you can take him home?

OMI: I know I'm spending an awful lot of time here with him, but since I . . . , my father saved his life and all, I figured he might need help with taking care of him.

NURSE: Are you sure it's not the handsome face?

OMI: Ha-ha., don't be silly! I mean . . . , he is handsome, but that's not it I just feel responsible for him somehow.

NURSE: I understand . . . ,whoops, they're paging me. Listen, I'll give you a couple more minutes with handsome, but that's it I'm afraid.

OMI: Thanks, but I'm leaving. I'm sure he's tired of listening to me babble on about nothing.

NURSE: All right . . . , see you tomorrow then?

OMI: It's a date!

In the mean time, special rescue forces and world military groups arrive at the impact area and set up bases of operations to get the investigation and massive clean up underway.

COMND GAULT: O.K. men, lets get this immediate area cordoned off and secured on the double! Sergeant, I'll need men stationed at these coordinates and they are not to be removed under my circumstances, understood?

SGT 1: Sir, yes sir!

COMND GAULT: I want everything ready and waiting for the colonel's inspection when he arrives at O eight hundred tomorrow, understood?

SGT 1: Sir, yes sir!!

COMND. GAULT: I will also need an escort to go into the restricted area with the rescue team to make sure they are safe and that nothing is removed, do I make myself absolutely clear Sergeant?

SGT1: Crystal clear sir!!

COMND. GAULT: Good, carry on soldier! Get me somebody that knows what the Hell is going on here on the phone soldier!

SOLDIER 1: Contacting central command as we speak sir.

COMND. GAULT: Splendid. Tell them to start sending the egg heads out here while the sun's still out I'm sure they're just dying to poke their noses and their million dollar equipment around out here.

SOLDIER 2: Sir! Reports just came in from our reconnaissance drone near zero impact area . . . ,it looks bad sir.

COMND. GAULT: That's one Hell of a skid mark! Is that what's left of our bogie soldier?

SOLDIER 2: As far as lean tell sir, yes it is. Would you like the drone to take a closer look. We're not going to be able to send in a manned team until that thing cools down. We're still getting a reading of a two to three hundred degree temperature difference coming from that thing sir.

COMND. GAULT: How long until we can send someone in?

SOLDIER 2: We estimate three days sir.

COMND. GAULT: Good . . . , that'll give me just enough time to get use to the idea of some egg head waltzing in and running my show. Where in the Hell is that damn Sergeant!? SARGENT!!

Back in Okinawa General Hospital, Omi visits the stranger yet again.

OMI: Papa, you came!!

AKIRA: Of course I did, brought some coffee too or did you have some already.

OMI: No dad, how considerate of you. Light and sweet?

AKIRA: Just like you.

OMI: Were there any other survivor at all.

AKIRA: Nope . . . ,it look like your friend here was the only one. Has he woken up yet?

OMI: No . . . , and I'm a little worried papa. What if he never wakes up?

AKIRA: Don't think like that sweetie. He'll wake up when the time is right. Right now, he needs you to have faith in that. You seem to like him don't you.

OMI: He's so different.

AKIRA: You're not lying . . . , I wonder what part of the world he comes from. He doesn't look like he is American. Maybe he's Polynesian or something like that.

OMI: I'm spending too much time around him, aren't I.

AKIRA: It's your time, spend it how you like.

OMI: What about business?

AKIRA: Well, we did get a little exposure over the accident, but I'm afraid two more crew members have left. They feel intimidated and threatened by the larger companies and I don't blame them one bit. Us mom and pop fishing industries are a thing of the past I guess.

OMI: Let them go, we don't need them anyway.

AKIRA: I'm afraid we do, honey. The less hands we have for the catch, the less fish we'll get.

OMI: I know you don't want to get into the conversation, but why don't we sell this business and try another . . .

AKIRA: Now sweetheart, you know how I feel about . . .

OMI: Wait a minute . . . , did he just move?

AKIRA: Look at the medical equipment . . . , it's going crazy!

OMI: He's moving!!! PAPA HE'S AWAKE!!!

AKIRA: Don't panic, I'll call a nurse!

OMI: NO!! Get me more COFFE!!

AKIRA: WHAT!!!

OMI: I meant him . . . , QUICK, he just drank mine and he's motioning for MORE!!!

AKIRA: That's CRAZY!! He just came out of a COMA and the first thing he asks for is COFFEE!?!

OMI: Just DO IT!! And give me yours until you get back with more . . . , Hurry UP!!!

The stranger devoured cup after cup of coffee until the doctors and nurses finally reached the room and ejected the two and their coffee.

NURSE: What's going on in here? What are you doing to that man!?

OMI: This man needs coffee!!

AKIRA: Hang on OMI. I'll get some more.

NURSE: The only thing you'll be getting is out of her . . . . , SECURITY!!!

DOCTOR: What's happening?!

NURSE: These two are trying to kill this poor old man by pouring piping hot coffee down his throat.

OMI: That's not true and you know it!!

DOCTOR: I'm sorry, but I'm going to have ask you to leave.

OMI: But she's LYING!!

AKIRA: Come on Omi before they call the cops.

SECURITY 1: Someone called for us . . . , Oh, Coffee, don't mind if I do.

NURSE: Hands off stupid, don't grab the coffee . . . grab those tow and throw them out for harassing a patient!

SECURITY 2: Yes ma'am!

OMI: Get of off me!!

AKIRA: Don't worry gentlemen, we'll leave in peace.

DOCTOR: See that they do.

NURSE: Poor man, look at him . . . he wants to try and kill them for what they did to him.

SECURITY 1: We'll be back to relieve you of the evidence . . . keep it warm please.

The stranger tried desperately, but in vain to chase after them and their coffee. They were told by the two officers that tampering with a patient could get them arrested so they'd better not return. Omi tried desperately to plead her cause, but to no avail.

OMI: Now what will we do? You could plainly see that the coffee was helping him!

AKIRA: Well if we go back we could be arrested.

OMI: I know, but there's got to be a way to get back in there and help him.

AKIRA: if you get yourself arrested your mother is going to do back flips in her grave, may God rest her soul.

OMI: Trust me daddy, I won't get caught. I'll pretend I'm a nurse or something . . . , they'll never find out.

AKIRA: Summersaults . . . , not back flips!

OMI: Stop it papa.

AKIRA: I may just join her if this keeps . . . , doing summersault that is.

Later on that evening during the change of shift at the hospital, the stranger snuck out of his room and managed to leave the hospital undetected. He knew that there would be no one on this small planet that could speak his language. That mean he would be all alone until he could find his companion that was with him. He needed to find him to continue his quest . . . , and have a decent cup of Ka-Fei. Oddly enough though, he could only see the face of the beautiful Omi. First pulling him from icy waters and then giving him a very, very weak substitute for Ka-Fei which gave him the energy to start his search. Her kind eyes and comforting thoughts were enough to allow him to believe she would help him further.

Just crossing the street was a task for him. The traffic seemed to be moving so fast that he thought he would never make it to the other side without being hit by one of the vehicles. He knew his only chance for survival was to get his hands on some Ka-Fei, but didn't know where to start looking for neither it nor OMI. He felt that the only logical place to start to look would be by the sea off he went, staggering into the darkness disoriented and weak.

On the way to the water front he tried to find and consume as much coffee as possible. He'd snatch a cup here and there while no one was looking for but the soothing effects of caffeine in Earth coffee was nothing compared to the Ka-Fei secreted by Ononot, the Living Bean. But he was afraid that he'd never find him. The escape pod crashed somewhere in the ocean awhile ago and without Ka-Fei, he would never be able to find it nor contact Ononot. He wandered for days on the end until the inevitable happened.

> MAN 1: I'd love to talk to you some more, but I know you're late for work.
> WOMAN 1: Don't worry about it. We still have time for another cup of coffee.
> MAN 1: Sure why not. Sir . . . two more cups of your delicious coffee.

As the waiter brought back the two steaming cups of dark roasted coffee, the fragrance was too much for the week stranger to try and resist. He pounces on the two cups like a panther.

> MAN 1: What th . . . ,?!!
> WOMAN 1: Ahhhh!!
> WOMAN 2: RUN!!! He's a crazy person!!
> WAITER: Quickly, someone call the police!! Tell them we've got a bum from the alley that's gone mad!!

Quickly the stranger swilled down the coffee and made a break for it down the nearest alley way. Meanwhile, back at the Yoshiuchi house, Omi was trying desperately to remove the stranger man from her mind.

> AKIRA: Still thinking about him huh?
> OMI: I don't know why. I mean, I kind of feel responsible for the guy.
> AKIRA: You know what they say, "once you save someone's life, you're for it." I guess it must be true.
> OMI: Do you think he's alright?
> AKIRA: Of course, he is in a hospital you know.
> OMI: Yeah, but doesn't mean they know what he needs.

AKIRA: And you do?

OMI: Never mind.

AKIRA: No Omi, I'm serious. What we did back there was stupid. We didn't know what we were doing to the guy, we could have hurt him or something . . .

OMI: Papa you know that's just not true! Didn't you see his eyes, he needed coffee!

AKIRA: No, I didn't get a chance to look, I was too busy fighting off hospital staff and security.

OMI: I don't want to argue about it dad, I just want to go and find out if he's okay.

AKIRA: if you get arrested . . . , I'm telling you!!

OMI: I'm not stupid dad, I'm not just going to waltz in there like nothing happened or something like that. I'm going to wait for a different shift to come on and disguise myself as . . .

AKIRA: If you want the family business just ask don't try to kill me by giving me a heart attack! Do what you feel you must do, but I'm telling you young lady, if you get caught don't look for this old man!!

OMI: I know you'll be there daddy, thank you!

AKIRA: Stop kissing me and hurry up!! Oh my sweet Tomizawa, I tried to raise our daughter correctly, but she just too head strong! What will I do? Disguising herself as an old woman, OMI managed to sneak into the hospital where the stranger was being kept.

OMI: Excuse me nurse, I'm interested in finding out where a certain patient was moved to.

NURSE: If you're talking about the weirdo that was in here couple of days ago, he's gone.

OMI: Gone! Where to?!

NURSE: Don't know. They said he just up and left. Good riddance I say.

He kept babbling on and on in some strange language about something. They had to strap him to the bed after some couple came in and tried to kill the guy by pouring hot . . . , hey where're you going . . . , I wasn't finished talking!! How rude!

Omi, in a fit a dismay, bolted down the hall and out of the hospital, searching in vain for the strange man who now could be anywhere. Three days had gone by and she had no idea where to start looking. She tried the bus and train stations, airport and finally the docks. There she overheard a strange conversation.

MAN1: This world's crazy man! Some derelict is running around swiping coffee man, coffee!

MAN2: Yeah, even from old ladies.

MAN1: I'm not lying man. Nothing is safe anymore. You can't even drink a cup of coffee without getting mugged for it!

MAN2: An old lady! When I heard about the I flipped my wig.

OMI: Excuse me sir, but when did this happen?

MAN 1: All this week. It's on the news.

MAN 2: Yeah, but the old lady thing happened this morning. I was there.

OMI: WHERE!?!

MAN 2: At the docks somewhere . . . , you a cop or something?

OMI: YES!

MAN 1: Man!! This guy must be bad news if they've got undercover cops after him.

Again Omi darts off, but this time she doubles back to say thank you. On the way there she passes a video store that had a large screen television for sale outside. She happened to catch one of the news broadcasting as she passed.

BROADCASTER 1: ". . . an unbelievable act of thievery today when a coffee grinder was forcibly removed from convenience store just hours ago."

BROADCASTER 2: "I don't know what's gotten into people these days, but it isn't good. There have been reports of this strange coffee snatching occurring all over downtown Okinawa and they seem to be increasingly in intensity."

BROADCASTER 1: "Here is a police sketch of the assailant. If you have any information on the where-abouts of the Coffee Snatcher, please contact us or your local police station for proper instructions."

BROADCASTER 2: "And please . . . , we can't stress this enough. If you are drinking coffee outdoors, please cover the top and disguise the cup, this man means business."

OMI: Oh my goodness! It's HIM!!

She continued her search and looked in every conceivable hiding place the docks had to offer, but with no luck. The Okinawa Docks were tremendous and it could take her days to find him. That is if he was still around. She also knew she had to find him before the police or she'd never get the chance to help him. She was about to give up when she noticed a very dirty man lying in a fetal position near an alley way. At that same moment a patrol car pulled up and noticed the same man. Although she couldn't see his face it had to be him because she saw the hospital name tag still around his wrist. She noticed the cops and acted quickly.

OMI: THERE HE GOES!! HE STOLE MY CUP OF COFFEE!!
OFFICER 1: Calm down ma'am . . . , what happened?
OMI: You didn't see that disgusting man steal my coffee!?!
OFFICER 2: What was he wearing ma'am?
OMI: Ahh . . . at shirt and some dirty jeans, please officers catch him
    before he gets away!! He went that way!
OFFICER 1: We're on it! Let's go get him!
OFFICER 2: We're gonna be heroes!!

The two officers sped off around the corner and left Omi alone with the strange man. She knelt over him and removed a thermos of the strongest coffee on the market, Colombia Espresso. The strange man trembled at the smell and like a drowning man reached out desperately for the thermos. He didn't even see who was holding it out to him until he had finished and collapsed in ecstasy. He finally looked up and saw his angel. The one that had been his side since he arrived on this world. He gave a sigh of relief and soon realized the worst was over. Finally, someone who understood. Omi called her father on her cell phone.

OMI: Hello, daddy . . .? Don't ask questions! I need you to come and
    get me.
AKIRA: I meant what I said about not helping you if . . .
OMI: I'm not in a place station I'm down by the docks with the man
    from the hospital.
AKIRA: So why can't you come on home . . .? Don't tell me you're
    afraid of that coffee snatcher guy they're talking about on . . .
OMI: The man from the hospital is the coffee snatching guy!! The
    police are looking for him as we speak and I've got to get him out
    of here before they come back!
AKIRA: He's the guy! You've got to get him out of the open befo . . .
OMI: THAT'S WHAT I JUST SAID!!!
AKIRA: Okay honey don't get excited, I'll be down there in a jiffy! But
    for now you've got to hide!
OMI: They're coming back!! I've got to go! Remember, I'm by the
    docks near the old freighter . . . , HURRY!'

Omi quickly searched for a place to hide the both of them before they were spotted by the police. The stranger was definitely not a small man and she struggled to get him on his feet and moving. They managed to get to the other end of the alley way before they were seen and made it to the old freighter.

> OMI: I know you don't understand me but you've got to get moving! I can't carry up this gang way and hold on to the railing at the same time! Oh . . . . , hurry up old man, you're taking too long!

The stranger seemed to understand and gathered enough energy to make his way to the deck of the old freighter. There he drank the remainder of the coffee while Omi called her father.

> OMI: Hello dad . . . .? Where ARE you!? You're taking too long! If the police arrest us it's going to be your fault.
> AKIRA: Now just calm down young lady and tell me where you're at.
> OMI: on the deck of the freighter, hurry up!
> AKIRA: I'll be there in a couple of minutes, but I can't drive too fast . . . , there are just too many cops out there.
> OMI: Great!! And they're all looking for US!!
> AKIRA: I brought some coffee with me and that's what took me so long.
> OMI: Good thinking I'm just about out. This guy's really something . . . . , he hasn't come for air yet!
> AKIRA: Stay low and keep the phone on. When I ring you, just start to come down, I'll be right around the corner.
> OMI: All right . . . , just hurry it up a little will ya? I wonder who you are. I've never seen anyone drink so much coffee and still be so weak . . . . , I'd be bouncing off of the ceiling by now.

Akira manages to get them without being noticed and they all head back to their house.

> OMI: You took too long!
> AKIRA: Be quiet and get this maniac to keep his head down before he gets us arrested.
> OMI: How's he going to drink!?
> AKIRA: I DON'T KNOW!?! THINK OF SOMETHING QUICK!!! COPS!!!
> OFFICER 1: Excuse me sir, we're looking for this man. Have you seen anyone who fits this description?

AKIRA: Ah, well . . . . no gentlemen, can't say that I have.

OFFICER 2: and the young lady?

OMI: No sir, ha . . . , not at all ha-ha!

OFFICER 1: Well if you do, stay away from him, he's dangerous.

OFFICER 2: And if you have any coffee, get rid of it NOW!! The guy we're looking for is crazy for the stuff . . . and he'll do anything to his hands on it!

BOTH AKIRA AND OMI: YES SIR!

Omi, realizing the officers smell the freshly brewed coffee, throws it out of the window and both her and her father try to keep the strange man from following it.

OFFICER 1: Who's that?

OMI: Oh ah, he's just my boyfriend . . . my father and I came down to the docks to pick him up . . .

Akira: Yeah, unfortunately he picked up some weird disease while he was away. It's a real mess. The other guys on the ship couldn't take the sight of him;;; started throwing up all over the place. The captain begged for us to get him off the ship because it started to affect his crew. Seems like the skin on your face just starts peeling away. See . . . , if you just look here you can see . . .

OFFICER 2: DON'T . . . do that thank you. Take him to a hospital or something. Let's go!

The two managed to get the stranger behind closed doors and try their best to communicate with him.

OMI: I'll try and talk to him while you go and make some more coffee.

AKIRA: We don't have too much coffee left, I think I'm going to have to go into town and buy some more. I also need to find our if anyone else found a stranger person floating in the water recently.

OMI: What is your name?

AKIRA: You're wasting your time, he can't understand you.

OMI: It doesn't mean I should stop trying. What . . . is . . . your . . . name?

GAIYYMANN: WWhatt . . . . , ees . . . , yonamie?

OMI: SEE!?! He does understand!!

AKIRA: Don't be silly. He's only repeating what you say.

OMI: it's a start. Pay no attention to him. Repeat after me . . . , I . . . am . . . Omi. Oh . . . me . . . he . . . is . . . Akira . . . Ah . . . kee . . . ra.

GAIYYMANN: I . . . am . . . , Gaiyymann.

OMI: Great! See he's learning!! He said his name was Gaiiyy-something.

AKIRA: Splendid. If you continue at this pace, you'll have ol' Gaiyy something ready for decent conversation in about five years.

OMI: You have no faith in me!

AKIRA: Sure I do. Look, he seems to be pretty smart. Why don't you use some of those "How to learn English/Japanese" tape I got you for school? That should help a little don't you think?

OMI: And we've got a lot of work to do . . . . , uh what did you say your name was again. Never mind, just watch these tapes and I'll get back to you a little later.

GAIYYMANN: Ka-Fei. Nat ell Ka-Fei

Omi: You mean coffee don't you? Cough . . . ee, not Ka-Fei.

GAIYYMANN: Caff . . . ieee?

OMI: Whatever.

The stranger continued to drink more and more cups of coffee knowing that soon he would be able to communicate physically with Ononot, The Living Bean and learn his condition and where-abouts. He felt he had let Ononot down, leaving him trapped in the sinking escape module. He concentrated fully on learning the language of the Pexians (humans) afraid that if he asked telephatically it may frighten them. So there he remained, seated in front of the television for hours on end, watching language videos and drinking cup after cup of coffee and espresso.

Meanwhile, back at the main impact area . . . ,

T.V. ANCHORMAN (America): "Ladies and gentleman the scene below is absolutely devastating. There's a damage as far as they eye can see and we cans see pretty far. The joint Military Operations Force composed of factions from around the world, have yet to comment on the history making event . . .

T.V. ANCHORWOMAN (Russia): ". . . and as far as I can tell whole villages have been wiped out. Trees and other foliage have been incinerated miles away from the impact area and reports of the explosion were made as far away as Barguzin near Lake Baykal, which is approximately two hundred and fifty-eight miles norht . . ."

T.V. ANCHORMAN (China): "The Chinese has offered the condolences to the families of the dead and will do everything within it's power to help get Mongolia back on it's feet. Rescue teams from the four corners of the world are still converging here with special equipment . . ."

T.V. ANCHOR WOMAN (Germany) ". . . can not or will not comment on why the stellar Defense grid did not function the way it was designed. But its highly unlikely that the families and loved ones of the thousands that died in this horrible tragedy will be comforted by any explanation given . . ."

T.V. ANCHORMAN (Japan) : "Not since Hiroshima has this reporter seen so much devastation in one place. Questions have been raised on whether or not this is the end or only beginning of a series as asteroids striking the Earth . . ."

T.V. ANCHOR WOMAN (France): "Millions, perhaps even billions of dollars in damages have been estimated in just the first day of the horrific event . . ."

T.V. ANCHORMAN (U.K.): ". . . the lips of the scientist that first arrived are sealed, possibly instructed to do so by the military. What does the military know that has lead them to keep the impact area such a well guarded secret? Attempts to fly near ground zero are met with Apache Attack Helicopters and Mig Twenty-nine Fulcrum jet fighters warning all approaching vehicles to turn around. No non-military vehicles are allowed within ten miles of the impact zone and those who do are forced either down from the sky or off of the road and immediately dealt with. Needless to say that there is something down there that the military feels strongly about keeping a secret. But just like it has always been, the truth will eventually surface. Hopefully it won't be as horrible as what has taken place here a few days ago in this quiet part of the globe. For B.B.C. News, this is Kenneth Williams."

And back at the main military compound one mile from the impact area . . .

COMND. GAULT: This doesn't make any sense, how could we be looking for bodies near or inside that mess when everything around it has been damn near incinerated?

LT. HUGGENS: That's our order sir. Straight from the top. Investigate wreckage for possible intact cadavers. It doesn't make sense to me either sir.

COMND. GAULT: They're treating it as if it were a plane crash or something.

DR. DELL: Maybe because that's what it is.

LT. HIGGENS: and who in the Hell are you civilian

DR. DELL: Dr. Harold Dell . . . , professor of Forensic Science at Cal. Tech. It's a pleasure to meet you Mr . . .?

COMND. GAULT: That's Lieutenant Higgens and I'm Commander Gault head of operations. Now what's this nonsense about this being a plane crash.

DR. DELL: Well . . . I'm the F.A.A.'s first pick every time a plane goes down in America. I really don't see the logic in my country calling me all the way out here for an asteroid.

LT. HIGGENS: Well what else could it be, I mean have you seen the size of that mess out there. I'm no expert, but even I can tell that's no airplane.

DR. DELL: Quite right my friend . . . , and yes, I've seen the wreckage. I don't think you know, but I was one of the ten men sent in to investigate the immediate area once the wreckage cooled down.

COMND. GAULT: We haven't been able to put together any pieces of this puzzle yet, but we will soon.

DR. DELL: Would you care to hear what I've come up with, Commander Gault?

COMND. GAULT: Frankly, I don't have the time doctor.

SGT 1: Commander! Reports on the bogie have been confirmed sir! I think you need to take a look at what they've come up with!

DR. DELL: I see that you are busy Commander, but this will only take . . .

COMND. GAULT: Sorry doctor, I'm going to have to ask you to leave for now, classified information an all that . . . . , I'm sure you understand.

DR. DELL: More than you realize Commander, more than you realize.

LT. HIGGENS: This way doctor. Follow me.

COMND. GAULT: All right soldier, what have we got?

SGT. 1: We've managed to put together a fuzzy picture of what we're dealing with here with the combined reports from The United Earth Forces Stellar Defense system, NORAD, N.A.S.A and a number of other reliable sources. Also the reports made from the first drone sent into ground zero four days earlier and the most recent investigations made by the joint science/military team sent in yesterday afternoon.

COMND. GAULT: Get to the point soldier!

SGT. 1: YES SIR! We are being lead to believe that the bogie was under intelligent control and after being damaged beyond its ability to maintain altitude attempted an emergency landing in the most unpopulated are possible . . .

COMND. GAULT: You mean to tell me that thing was being piloted!?

SGT. 1: That was the conclusion made based on the given date from . . .

COMND. GAULT: Intelligently controlled by whom, Sergeant?

SGT 1: We still don't know exactly sir, but no country has stepped forward to claim they are the ones behind it and from what intelligence has gathered, even if they do they would be lying . . .

COMND. GAULT: Of course they won't! Who in their right minds would want to claim responsibility for this mess . . . Why would they be lying Sergeant?

SGT 1: Because for one sir, the technology retrieved from inside the wreckage is unknown to all f the scientist gathered here.

COMND. GAULT: That can be ruled away as a cover-up from one of the countries represented in then science crew.

SGT 1: There's something else sir.

COMND. GAULT: Speak.

SGT. 1: Bodies have been recovered sir, non-human bodies.

COMND. GAULT: Non-human Sergeant?

SGT. 1: See for yourself sir.

What the Commanders saw shocked him. Thirty tall, strangely armor-clad bodies were laid out in size order for visible inspection by medical and scientific staff.

The body sizes ranged from six to eight feet tall and better. What amazed him the most was that some of the bodies were still intact, even after the impact and explosion. Strange implements were also found at their sides which appeared to be weapons were undamaged.

COMND. GAULT: How many others have seen this Sergeant?

SGT. 1: Just this campy sir, and the doctor you were speaking with a moment ago.

COMND. GAULT: Find Lieutenant Higgens and have him bring back the good doctor. I think I'm going to need to finish the conversation we were having.

SGT. 1: Sir, yes sir!

COMND. GAULT: And while you're at it soldier, see that no one enters or leaves this site without being debriefed by me. Is that perfectly clear Sergeant?

SGT. 1: Done sir!

CHAPTER 2

Communication Factor

Dr. Cocoa: They're not going to blame this on me. I'm not going to tell them!

Wife: They couldn't if they tried! How dare they, as much time and effort you put in this project . . . it almost caused our divorces for goodness sakes!

Dr. Cocoa: I want to make sure that whatever happens to me, you and the children will be sage. You've to promise me you'll go back to the island things go bad for me . . .

Wife: But I'm your wife . . . I . . . I can't just leave you, abandon you! No matter how bad things get!

Dr. Cocoa: Think about the children Louise! Now you can't go off thinking . . .

(missing items)page 25 pdf.

Wife: I know my love

Dr. Cocoa: And don't you worry about protecting myself or even the children. I'll see that nothing happens to you . . . I promise.

Meanwhile, back at the crash site, special equipment and transport housings have been set for the removal of the alien remains and debris from what is left of the spacecraft. The military has set up security rings around the site, 5 to be exact. The closer you get to the site, the higher the security.

Commander Gault: Lt. what's the status on the remains? We can't keep them out forever they've got to be starting to show some decay by now correct?

Lt. Higgins: negative sir. No degeneration in tissue so far as we can tell. We're not sure, but we think it's the armor they're wearing. Each one is giving off some sort of stranger energy field. We don't know what it's doing or even what kind of energy it is, but what we do know is that it is very cold out today and the temperature of those bodies don't go below 108°.

Commander Gault: So their armors got built in heaters so what?

Lt. Higgins: I don't think you understand sir. The outside of that armor is reading temperature of only 31.7° F sir; it's the tissue temperature that's 108°.

Commander Gault: Those things aren't still alive are they?

Dr. Dell: If they are, it isn't life as we would know it, which is why we need to remove these bodies for further analysis.

Commander Gault: We're working on that doctor, we're working on that.

Dr. Dell: it's been more than a week!

Lt. Higgins: These things take time sir. We don't even have a facility large enough to take these things even if we could move them now. No one planned for this to happen.

Commander Gault: At ease Lt., the good doctor here knows exactly what we are and arent capable of.

Dr. Dell: That doesn't change the fact that valuable time isn't being lost . . . not to mention being exposed to whatever energy that thing is putting out. The engines on that thing are still intact you know . . . goodness knows what it's leaking into the atmosphere!

Commander Gault: We're taken all precautionary steps to insure the safety of everyone here, Hell . . . . you think if I thought it wasn't sage I'd still be here!

Sgt. #1: Lt . . . . , sir we need you to come see this sir.

Lt. Higgins: See what Sgt?

Sgt. # 1: It would be easier if you just came sir. Something is happening to one of the soldiers.

It's been 10 days since Omi found Gaiyymann and had been nursing him back to health with cups of fresh coffee and the occasional piece of raw fish to Akira's dismay. Gaiyymann amassed large amounts of information about where he was and soon knew that it would be time for him to leave in search of Ononot. He knew that Ononot was where they had lost parted but was not strong enough to reach him.

Akira: How's our friend doing?

Omi: if you consider drinking cup after cup of coffee and eating raw fish normal, then he's doing great!

Akira: and his conversatin?

Omi: Not much I'm afraid. He seems to be more interested in watching T.V. than talking but I haven't given up hope yet. He'll come around.

Akira: You think that crash did something to him . . . you know mentally?

Omi: I don't know. He seems to be really interested about it on T.V., although I doubt he understands what he's watching too much.

Akira: Don't be surprised. He has been going through those language tapes pretty quickly . . .; I wouldn't be surprised if he could understand what we we're saying right now.

Omi: You're being silly dad, those tapes translate Japanese to English.

Akira: And vice-versa . . .; I suppose your right. I mean, he doesn't speak either language, I don't know what makes any of us think he'll understand us. You did try to speak to him in English right.

Omi: yeah at the hospital. He doesn't speak it, or understand it; if he did we'd be speaking to each other by now. We do know one thing though . . .

Akira: and that would be?

Omi: He understands what we don't understand and he doesn't understand what anybody is saying to each other.

Akira: What did you say?

Omi: See what I mean?

Gaiyymann smiled. He did understand, but he couldn't let them know that, not yet. He had to keep that and many other things a secret. How was he going to get to Ononot? And where would they go once they were together? This place. It had to be here. There is no other place! He felt oddly safe her and knew Omi's intentions were good. But for now he would have to use her, and time was short.

The next day at the private docks at the Yoshiuchi Fish factory, Gaiyymann decided that now was the time to start to show some progress in understanding the language and use the opportunity of the Fish Factories' short staff to gain a spot on one of the boats and try to make it back to the spot where the pod went down. When no one was around, he donned a uniform and snuck aboard one of the ships.

Akira: Where's your friend? At home again staring at the T.V. again or did you bring him along for another breath of fresh air? You know that taking him out can be dangerous.

Omi: He can't stay at home forever, and besides, he seems to make himself a little useful around here.

Akira: How, by eating all the fish we catch?

Seamen #1 : We're casting off sir! It's fishing time!

Akira: Very well. Take us out. Omi, stay topside with the Yule man and firstmate, I've got to check the nets one more time.

Omi: O.K> anyway I left him at the office he'll be fine there . . .

Akira: You did WHAT!? Are you crazy!? Half od the police department is looking for this guy and you decided to leave him there!? What if someone finds out!? We'll be arrested for harboring a criminal!!

Omi: Don't worry! Everybody here already knows he's here because I told them what happened and asked for their help. They all remember me saving the guy . . .

Seaman #1: That's right she was fantastic,

Seaman#2: Yeah, I'd have never done that, that water's freezing

Akira: Omi, you are far too trusting! Don't let this stranger be the death of poor father.

As the ships pulls off towards their fish grounds Gaiyymann proceeded to the front to try and get a fix on Ononoy physically. He was weak but it was working. The pod was still in the fishing grounds and Ononot was still alive. The only thing left was to get Ononot.

Seamen#1: Hey, you shouldn't be on board! How did you get a uniform! If Omi or the old man finds you snuck on board, their gonna kill me. Listen, don't know if you can understand me but you're gonna have to go below until we get back. Let's go.

Gaiyymann nods his head and follows him to the galley.

Seamen#1: puts a cup of coffee in front of him

Seamen#1: Here, drink this and don't move, I'll have to find a place for you to hide. You're gonna get me fired! Work is hard to find for fishermen these days you know.

As the seamen left to find a place to stash Gaiyymann, he snuck up to the bridge and physically influenced the navigator to change course slightly and head for the pod!

Back on board the flagship The Deep Neptune . . . .

Seamen#1: Sir, I think the coral is off course, look at her heading.

Akira: I think you right Kenji. Radio them and tell them to correct their heading.

Seamen#1: Yes sir. Fishing vessel the coral, fishing vessel The Coral this is the Deep Neptune do you copy . . . over?

Akira: Mamorn, make yourself useful and ready the nets we're almost at the drop point. After that go down below and prep up the freezers for cold storage, take Hiroki and OSmau with you.

Seamen#2: You're working me to death!

Akira: Any luck with The Coral Kenji?

Seamen#1: They say they're on course sir.

Akira: Give me the radio! Coral do you read. This is Captain Akira Yoshiuchi of The Deep Neptune! You're off course change your heading immediately!

The coral: We are on course sit these are the coordinates you gave us at this mornings de-briefing . . . longitude 37 latitude 131 heading due west at 21 knouts . . . over.

Akira: Kenji give me the map . . . Mamoru, you still here, get moving! Coral Deep Neptune to Coral. You're a half-mile off course you're not reading your compass right! Make adjustments . . . 15 degrees north. Kenji, at there speed where should they meet up with this?

Seamen#1: At . . . about . . . 1 quarter mile away from the drop point sir. Hey, it looks like they're heading the for the same spot where we picked up that weird guy last week.

Akira: Not, important, we need to concentrate on fishing. Maintain contact with The Coral, I;m going below to check on Mamuro, the lazy bum!

Back at the crash site, a meeting has been held by Commander Gault and his staff along with the scientists to discuss the stranger occurrence with one of the soldiers.

Commander Gault: Good morning gentlemen as we all know by now, we are not dealing with an ordinary accident. A few days ago you were kept in the dark about what we have here. Only a few of you were given the clearance for what I'm about to tell you now. But circumstances have made it necessary to explain to those that don't know just what we think we're dealing with here. As you all know an object struck this planet on April 3rd 2032 at approximately 0458 hrs E.S.T. which cause almost insurmountable damage to the country of Mongolia and a poisonous cloud that has half the country evacuated to nearby China. What all of you don't know is that we believe that this object was not an asteroid but vehicle of other worldly construction piloted but extraterrestrials of a military nature. 30 large humanoid bodies and remains were found near the crash site. Most of them intact, and some form of protective armor. They also looked as if to be carrying weapons of some sort, we haven't confirm that. We have not been able to enter the vessel due to the extreme heart of the outer shell at first and then the strange energy readings we were getting from the bodies. We now have a problem. It seems that prolonged exposure to the crash site presumably within the radius of 50 feet or so will cause the human body to, put it layman's terms . . . . burst into flames from the inside out.

Scientist#1: Spontaneous combustion that till hasn't been proven to be a fact.

Scientist#2: It is quite possible, under the right circumstances . . .

Scientist#3: It's 40 degrees out there! What circumstances!

Scientist#2: We don't really know what causes it to happen . . .

Scientist#1: We do know, that thing out there . . .!

Scientist#4: Are we sure we're safe? I mean to say that we really don't know what we're dealing with here . . .

Scientist#3: And why did you wait almost 2 weeks to tell us all of this Commander!?

Commander Gault: Simply because you didn't need to know! Now gentlemen may I continue?

Scientist#4: I'm going to get myself checked out immediately!

COM Gaul: Now . . . as you see in this picture the soldier body was not completely consumed by the flames his limbs and head were left intact. This happened while on break inside one of the wreck rooms. We believe that due to the tissue damage, the response time of the fire team reacting to the fire alarm getting to the fire and putting it out and the absence of any accelerants aside from the soldier's own clothing and his side arm in which all of the rounds went off, simultaneously as ballistic reports have confirmed that the body of the soldier reached a temperature of 1000° F in a matter of only 2 minutes maybe even less.

Scientist#3: That's impossible! Where are the reports? I need to see them right now!

Scientist#1: Unbelievable!

Commander Gault: Obviously, we can't afford this incident to leave this area, nor can we allow any of this information to leak out as well. Whoever has ring 1 clearance are now hear by contained inside of ring 1 until this can all be can be sorted out. No one, not even myself is cleared to leave this area until we can figure out what happened to that soldier and if it will happen to anyone else.

Scientist#4: You can't do that! You can't hold us hostage here!

Scientist#2: I'm afraid he can Dr. Gigot, have you forgotten the confidentiality forms you signed?

Scientist#4: Are you crazy? This has nothing to do with confidentiality!

Com Gaul: I'm afraid that you're wrong Doctor. We can't afford for this to get out until we understand what we are dealing with . . .

Scientist#4: I don't care! You can't force me to stay!

Scientist#1: And if they allow us to leave, what will you do then, burst into flames in your living room or better yet spread this . . . whatever it is to your family or loved ones? We don't understand what we have here but I guarantee you if you do leave, if any of us leave, we won't find anyone outside of this room who will! Our best chance is to remain, claim and work together! We've all spent sometime working near the crash site . . . you aren't the only one that feels threatened here!

Scientist#2: Dr. Hall sateen is correct, I'll have to agree with him . . . please sit back and be rational!

Commander Gault: Gentlemen please! We're wasting valuable time here. The body is being examined as we speak for possible contaminates but we don't think that's the case. We believe prolonged exposure to the wreckage and the bodies' chemistry are the culprits.

So how are we going to examine this thing without bursting into flames!

Commander Gault: We're working on that problem as well, Lt. Higgins.

Lt. Higgins: Thank you sir. We believe some form of unknown radiation suit strong cause. Now this particular soldier was not equipped with a radiation suit strong enough to block whatever energy this things is emitting. New deep space suits are being sent in to deal with this particular situation but we can only send in 8 people because only 8 were ever made.

Scientist#1: Now all you need are 8 volunteers . . .

Scientist#4: Good luck with that!

Com Gaul: Now gentleman! I'm not going to stand here and beg for volunteers, we just need one of you to go the rest will be military . . .

Dr. Dell: I'll go.

Lt. Higgins: I'm also on the list. Gentleman need I remind you that although the risk is great, so are the rewards! Think about it stepping into a craft from another world it's historical.

Scientist#2: All right! You don't have to sell the idea to us! We all know how significant a discovery this is.

Scientist#1: Dare I say I'm tad interested myself, how long will we be inside that thing?

Lt. Higgins: Not long, but long enough to get a decent look around.

Scientist#2: And how safe are these suits, have they been tested?

Commander Gault: Yes . . . because of this suit. It was badly damaged but he lived. He's still in service today despite the incident.

Scientist#4: I'm not in my 30's you know!

Iredell: I already said I would go. I'm the only one here qualified for this anyway.

Lt. Higgins: Thank you doctor, be prepared to make history. For the rest of you, samples of the expired soldier and his surrounding environment have been provided to you for your examination. Thank you gentlemen that will be all.

Meanwhile, back on the board The Coral, Gaiyymann sensed he was drawing near to Ononot and figured the waters would be deep. He needed to find a diving apparatus to aid in his search. He makes his way to the desk of the ship towards the rear putting on the equipment as he goes. He saw one of the net operators and physically influenced him to drop the net. Gaiyymann knew he was getting weaker every time he used his abilities but he had no choice. Just then he noticed two fast moving boats heading in their direction from in front apparently to intercept The Coral. He also noticed a larger military ship not far away off the starboard side.

Seamen#2: We've got vessel contact off the starboard bow! They're trying to cut us off!

Seamen#3: Is that the military?

Seaman#4: Shut up! I'm trying to talk to Captain Yoshiuchi! . . . I'm sorry sir, the navigator made a mistake in bearing, the new course has been set but I think we're being haled by tow ships . . . we don't know who they are . . .

Seaman#3: They got guns. Big ones! It's got to be the military!

Seamna#2: Maybe they're pirates

Seaman#4: Don't be stupid!! Can't you see that big ship out there, it probably is the military. Captainb Yoshiuchi I think the military wants us to stand down and prepared to be boarded.

Seaman#2: It's the army! They're telling is to cut off our engines and stand by! I hear them over their loud speakers! What do we do?

Seaman#3: What do you think stupid . . . stop the boat or they'll sink us!!

Seaman#2: I don't think they'll sink us, but I am stopping the boat.

Captain, Yoshiuchi the military wants to board us sir, we have to stop. We're going to be a little later at the drop point.

Seaman#4: We're not at the drop point?

Seaman#2: Stop all engines stop. Attention all hands prepare to be boarded. What are you talking about?

Seaman#4: Well I thought we were here since the net was dropped at all . . .

Seaman#2: I never gave orders for that! What in the world I going on around here!

Before anyone could notice, Gaiymann was over the side and into the water. He closed his eyes and let the psychic connection between him and Ononot guide him through the depths. Slowly he descended through the calm darkness of the ocean. He was no stranger to swimming and enjoyed his swims on The Planet Kingdom Blendeen but this was different. This time he was trying to save a friend. Back on the Deep Neptune . . .

OMI: Dad what's going on? Where do the Coral go . . . we can't see her?

AKIRA: The military is boarding her. Kenji just told me they must have wandered into some sort of military operation . . .

Omi: In our fishing area? Why weren't we notified earlier? That doesn't make any sense . . .

Akira: A lot of things are starting not to make, sense, Mamuro . . . prepare the nets we're almost at the drop point!

Omi: Daddy . . .! The military is boarding one of our ships! Didn't you hear me!? We weren't notified of anything! What are you going to do!?

Akira: What we came here to do . . . fish.

Omi: Stop being stubborn! O.K. you fish, I'm going to find out what's going on!

Omi heads to the bridge and hears the voice of her father blare over the loud speakers to drop nets.

> Omi: Kenji, what's the status on The Coral?
>
> Kenji: She not answering, Omi, I don't think the military is letting her respond.
>
> Omi: That's just great. Go see if you can lend a hand down on deck. I'll stay here and keep trying to raise The Coral?
>
> Omi: Yeoman, cut to trawling speed and circle in a 15° degree drift towards the direction of the Coral be ready on my command to head in that direction at full speed on my order . . .
>
> Yeoman: I don't think your father . . .
>
> Omi: My father's busy, and if he's not on the bridge you do what I say, got it!?
>
> Yeoman: Yes Ma'am!

> The nets are cast off the Deep Neptune and the Rising Sun and it is business as usual for Akira and his crew, although fishing is not exactly on the minds of everyone, especially the crew of the Coral . . .
>
> Soldier#1: Who is the captain here?
>
> Seaman#1: I am. What seems to be the trouble here?
>
> Soldier#1: The trouble is that you're impending on a military operation. Turn your vessel around and leave the area at one.
>
> Seaman#1: We're just fishing here, nothing more. We've been fishing these waters for . . .
>
> Soldier#1: I don't want excuses and you don't want problems so do as I say and turn around.
>
> Seaman#4: They'll have to give us time to pull in the net, we just dropped it.
>
> Seaman#2: Yeah and if we stall 'em long enough we might even be able to snag a net full of fish. The old man will forgive us.
>
> Seaman#1: Sounds good but we can ignore the army. If they say leave, we leave . . . you guys find out who dropped the nets and have them pulled back up. I'll go to the bridge and give the order to cast off!
>
> Seaman#2: Man what a wasted day this turned out to be!
>
> Seaman#4: And a crazy one too! You take care of the nets and I'll find out who dropped them.
>
> Meanwhile Gaiyymann hit the bottom, it was dark and cold but he continued on. The pressure was squeezing his lungs but held on the until he felt the smooth metal of the capsule. It was undamaged

and slightly embedded in the ocean floor but that didn't slow him down. All he had to do was get the net to engulf if the capsule and Ononot would be safe. He had to act fast, he was running out of strength. He used up his last remaining ounces of strength to free the capsule from the ocean floor and push it in the direction of the trading nets.

Seamna#2: This is the coral to the fishing vessel the deep Neptune, do you copy, over . . . . We're resuming re-calculated course and speed to rendezvous with, you and the rising sun . . . E.T.A . . . 30 minutes. Sorry about the delay, we will have an explanation for you when we meet . . . over.

Seaman#2: We've got something in the net, it's heavy as hell! Looks like we hit the mother load!! Someone get on the horn and tell Tamaki to ease off the throttle and find out what we got!

Seaman#3: Hey Boss! We got a catch . . . I repeat we got a catch! How big is the blip. Did we net a school or what! The net is showing 3 ½ TON pull. That's a crap load of fish!

Seaman#1: Navigator . . . what's the situation on sonar, did we net a school of tuna or something!

Navigator: I think "or something" would better explain it. I got a small blip on sonar in our nets but unless that thing is made of lead or some other kind of metal there's way something that small could weigh so much!

Seaman#1: Could it be another ships anchor?

Navigator: Too small, it's the size of a man.

Seaman#5: Maybe it's a golden statue or something, to hell with fish, we could be rich!

Seaman#1: Whatever it is as long as the nets aren't snagged, we're getting out of here. Maintain speed and course. Attention on deck, continue to pull up the nets slowly, we'll check out what we have in 'em once we rendezvous with the other ships.

Seaman#3: The boss says keep pulling it up slowly! We'll check 'em when we link up with the rest of the fleet!

Seaman#4:Right! Now listen, I want all hands on deck immediately. I need to find out who dropped those nets and why!

As the crew of the coral gathers on the deck, Seaman #1 remembered Gaiyymann! In all of the confusion he had forgotten about him! But he hadn't seen him since the galley and no one else on board reported seeing him. Maybe he found his own hiding place. Not the time to worry about that now, he had to make up a legitimate excuse to use for being so far off course and having an unknown object in his nets.

Seaman#4: I've only got a hand count of 19, were short one man . . . We've got the Captain, 1st mate, Navigator, Tec guy the 3 mechanics, the Yeoman and Both divers, the cook all 4 net ops and 4 storage personal . . . hey, where's the medic . . . you don't think . . . oh no, man over board . . . man over board!!

Seaman#3: Holy crap. Boss!! We've got a man over board!! I repeat, man over board! !

Seaman#1 : You said the thing in the net was as big as a man . . . Oh NO NO NO!! GET THE DIVERS IN THE WATER NOW! Ah, Uh ALL ENGINES STOP!!!

Navigator: Should we radio for help! The military is right there!

Seaman#1: Uh no, NO!! We'll handle this ourselves! Just keep your eyes and ears open; I'm going on deck!

Gaiyymann managed to get tangled up in the net as well as the pod.

As he ran low on air he let the cold of the sea take its toll on his body. Even as strong as he was, he was no match for the deep. He had to hang on. They had to pull the nets up sometime and soon he and his friend would be united once more. He closed his eyes and embraced the silence and the nothingness around him. He felt horrible not to know whether not this would be the end for him and yet another planet enslaved by the hypnotizing decadence of ka-fey. He felt weaker and weaker as his oxygen ran out. Only a matter of time now. If he only had more time. Where was the beautiful human female that had saved him before. He would never see her again, never be able to tell her how thankful he was, never see his home world again he would die here, at the bottom of an ocean on a planet eons away from his and no one would care Yeoman: Omi! The coral has stopped moving and there's no response from the bridge!

Omi: All right that tears it . . . this was a lousy day to go fishing anyway . . . prepare to go underway! Retract the nets . . . full speed ahead!

Seaman#2: Hey! We finished already old man?

Kenji Capt., your daughters giving the order to cast off

Akira: Belay that order! I'm going up, keep fishing!

Navigator: It's official; she's dead in the water, still no response from the bridge, although I'm getting a strange reading form their net sensors. It's not a snag but they did net something heavy! 3 Y:z tons to be precise but nothing showing up on sonar array, at least not theirs any way, except a small blip. Too small to be reading right.

Omi: Are we too far away to use ours?

Navigator: Yeah, we need to get closer.

Akira: Omi!! What are you doing!

Omi: The Corals in same sort of trouble dad. We've got to get to them, they're dead in the water and there's no response from the bridge!

Yeoman: There's also something in their nets, although their sonar array seems to be out the net sensors are picking up one hell of a drag!

Navigator: Could be a school of tuna!

Akira: At 3 12 tons!! That's a nice catch! How is the rising sun doing?

Yeoman: Not so good, If there's tuna over their, imagine the catch at 3 Y:z tons each ship!

Navigator: Maybe more! Hot damn, we'd better get over there before they get away!

Akira: Maybe your right! Draw the nets and full speed ahead! Give the same order to the rising sun lets see what we got! Great idea Omit

Navigator: Yup, all her idea . . .

Yeoman: See, I told you your father wouldn't mind!

Omi: Uh . . . . yeah, thanks!

As the two ships prepare to cast off, the divers of the Coral reach the near dead Gaiyymann, limp and tangled in the heavy nets.

Seaman#3: It's confirmed it's our missing man! Judging from the time he's been down there and his lack of response. I don't think he made it boss. They say he has diving equipment

seaman#1:He might still be alive then just unconscious

Seaman#3: Should I give the order to cut the net

Seaman#1: Of course cut the net! Get him out of there!

They bring Gaiyymann's lifeless body to the surface and attempt C.P.R He coughs up fluid but does not regain consciousness. They take him down to the med lab.

Diver #1: There's something else down there.

Seaman #1: What? Fish?

Diver #2: No, we don't know what it is. It's as big as a mini van.

Diver #1: Bigger, bigger than that. I can't tell what is was but it wasn't a fish.

Seaman #3: Should I have the guys finish pulling up the net?

Seaman #4: No there's no use. All I was concerned about was our man.

Seaman #2: And how and why did the medic don half a divers suit to go swimming in this water?

Seaman# 1: I don't know but I do know that that wasn't our medical officer that they just took down to the med lab.

Seaman #3: It was who then?

Seaman#1: Someone that's going to get me fired.

Seaman#2: Don't worry, this wasn't your fault, Capt, Yoshiuchi wouldn't fire you, he needs you.

Seaman#3: He needs all of us including that dead guy. Hey let's go see if he is really dead

Seaman#2: He just coughed. Dead men don't cough

Seaman#1: I'd like to know what's in those nets!

The three fishing trawlers meet up and information and stories were exchanged by both the crew of the Deep Neptune and the Coral. Akira and Omi stand over Gaiyymann's lifeless body in the moo lab.

Omi: I don't freaking believe it.

Akira: This guy gets around huh!

Omi: He must love the water

Akira: Yeah but he must love drowning even more.

Kenji: Are we going to keep staring at his body?

Both: Omi and Akira look at Kenji; Kenji looks down to the floor.

Omi: I've got it! It worked before, lets try it again! Kenji, get me some coffee!

Kenji: So . . . . we are just going to sit have and stare . . . .

Both Omi and Akira scream at Kenji to go and get coffee.

Akira: What about that thing in my nets, you think he knows what it is.

Omi: Remember I told you the Coral was headed for the first place we found him? And remember how he was holding onto something.

Akira: Yeah, it kind of looked like he didn't want to let it go, probably didn't want to drown.

Omi: Or maybe didn't want it to sink . . .

Akira: I really have a strange felling about this guy but our paths are always crossing, weird huh?

Omi: You think mama knows what's going on?

Akira: Probably, you know . . . she has always been good at keeping secrets.

Kenji: Here's your coffee . . . man, this stuff smells strong!

Just then, movement from Gaiyymann. Omi poured the coffee into his mouth and waited. Gaiyymann gasped and reached out for more. Ahhh sweet Ka-fey! How he longed for it, he reached out.

Omi: He's coming too! Man, this guy is something else!

Akira: I've never seen anything like him! He's a real survivor!

Kenji: Looks like he loves his coffee too!

The three ships returned to port, all nets almost empty except for one. The strange object glistened in the midday sun inside of one.

Back at the crash site, Commander Gault and Lt. Higgins prep up the space suits for craft entry tomorrow. Security around Ring #1 has been nearly tripled. All outside communications been terminated and traffic has been frozen. Only supplies authorized solely by the Commander are allowed movement! The suits are so new and experimental they have to be monitored by N.A.S.A. techs only and those that designed them.

Sgt#1 Sir the deep space suit techs want to have a word with you sir they're outside now.

Lt. Higgins: Send them in Sgt.

Sgt.#1 Yes sir.

Teach #1: good afternoon Lt., my name is Dr. Rampashard and this is Mr. Billings 1st technical advisor, Mrs. Gathings Systems Monitoring Specialist and Dr. Liu, Cosmic Radiologist.

Lt. Higgins: Good afternoon Gentlemen, and Lady how many I be of assistance?

Tec#1: We hear that you will be one of the group venturing into the object, correct?

Lt. Higgins: Correct, The commander needs one of his closest officers on the scene first hand . . . just in case of any unforeseen occurrences. You understand I hope.

Tech#2: Yes, we understand this is still a military operation.

Lt. Higgins: Good. Personally, I can't wait to turn all of this over to the eggheads . . . present company excluded of course,

Tech#4: Not to be facetious but aren't you little young to be a commander?

Lt. Higgins: Well, the space defense grid is quite new and the technology is pretty advanced. Not too many people are trained in it. I myself was personally trained by its chief designer Dr. Kaneohe Cocoa.

Tech#3: 1guess that makes you kind o fan egghead as well, does it not?

Lt. Higgins: Ah . . . Touche, Mrs. Gathings, very well done. Yes I would suppose I'm not the young military 'grant you were expecting.

Tech#1: Word is he's in big trouble because of this.

Tech#2: Poor guy, they're going to be looking for someone to place blame

Lt. Higgins: This is very true, and very unfortunately for him, none of what's going on here can ever be released.

Teeh#3: Would that matter to him. In what way if so?

Lt. Higgins: We're getting off of the subject. Like I said before, my military and space knowledge make me a needed addition to this mission. If anything out of the ordinary happens in there I guess I'll be the go to guy.

Tech#1: This is truly a historic occasion, unfortunately none of us will be going with you, I mean we will be monitoring all the equipment from the base so technically speaking we'll be right over your shoulder.

Tech#4: We're also aware that we won't be leaving Ring#1 anytime soon. Would you mind explaining why?

Lt. Higgins: We've . . . been having certain problems with security here sooo . . . extra precautionary steps had to be taken.

Tech#3: A problem with security? Here? Well I find that kind of hard to swallow wouldn't you agree David?

Tech#1: Nothing is completely airtight when it comes to the government Silvia.

Lt. Higgins: Well, what brings you by my office, aside from curiosity of course.

Tech#3: Of course.

Tech#1: We just need to know if there were any . . . special concerns dealing with this object, things that would need to be brought to our attention?

Lt. Higgins: You mean. N.A.S.A.'s equipment

Commander Gault: He means the government's equipment!

Lt. Higgins: Yes, sir . . . . excuse me sir; I didn't see you come in sir.

Com Gaul: At ease Lt. so, these are the N.A.S.A. guys?

Lt. Higgins: Yes, sir this is Dr . . . .

Com. Gaul: No need for introductions, I know who they are.

Tech#1: You must be the Commander! We were just about to ask about any particular problems with the project we should know about.

Com. Gaul: No and if there was it's not for you to worry about, that's what

Lt. Higgins gets paid for.

We need you concentrate on what you were sent here for, to get my personnel in and out of that object safely. Now if that was all you had to ask. I'm going to have to ask you to join the other staff in staging area 10 for more debriefing, thank you gentlemen. Lt Higgins a word alone please.

Tech#3: And lady.

The techs were escorted out of the office by military police and driven to staging area 10 the point of effect by the strange energy given off by that craft. Later that afternoon the meeting. between Dr. Cocoa and his staff is about to take place.

Dr. Cocoa: O.k. we all know why we're here, we've got to get our story straight before we go in front of the board again. All though we've done nothing wrong we still need to prepare for some grueling questions. Now we're all well versed in our fields of expertise but we have to be on the same page . . . .

Dr. Wong: I still say we need a lawyer!

Dr. Cocoa: No lawyers, how many times do I have to say that! No one is handing anything . . . we've all done no wrong here!

Gregory: All here Kaneohe, we all know that! But we also know how devious the governments of the world can be.

John: He's right. We need to know everything they know about this.

Dr. Cocoa: You know that's not going to happen, but some info will help, John, what have you got?

John: Well my connections tell me they are gearing up to name a scapegoat, probably one of us.

Dr. Cocoa: And since it's my baby, it will probably be me. But I have my connections as well and they tell me they're going to have a tough time trying to prove negligence on any ones part.

Dr. Wong: But we still have to know why it failed in the first place.

Dr. Cocoa: It didn't fail! That grid was fail-safe. Nothing we could think of could have passed its defenses! You know that! We simulated every feasible occurrence into the sim com and every scenario was handled. If the governments thought there would be a problem they would have never installed the blasted thing.

John: He's right, they spent billions upon billions to build it and countless hours of training time were spent on manpower training.

Dr. Wong: That's why they are going to need a scapegoat, or scapegoats! I'm telling you we need legal defense.

Gregory: Please!! Enough with the lawyers already!

Dr. Cocoa: John, are you sure your contacts can be trusted!

Dr. Wong: You can't trust anyone nowadays!

John: Yeah, they can be trusted. I know one personally.

Dr. Cocoa: Good, the same goes with me. Now, me all have our material with us, correct? We need to exchange notes and really know each other's role in development well enough to beach each other up should there be a need to do so. We will only have a few weeks before they call us up in front of the board again, so prepare well.

Dr. Wong: Gentleman, if I may. I'm sure I speak for all of us when I say we are all friends here. I've known John personally for years and I do trust you all. What I'm having trouble here with is the lack of knowledge we have of what actually hit the ground there in Mongolia. I for one would like to personally go there and see that thing.

John: There's nothing there except a huge hole in the ground.

Dr. Wong: That's what the news was told to report, you know that.

Dr. Cocoa: I'm going to go to the impact zone and talk to one of my contacts. I've been told that he had been selected to head one of the investigative teams there. I'm sure he'll give me clearance. The rest of you prep up, we'll have another meeting when I return. Gentlemen I am sure you we will not to be held responsible for this . . . that is all

CHAPTER 3

DARK PATHS

The mid-day sun beams down on the crew of the Deep Neptune as they stare aimlessly at the strange objects in the ship's fishing nets. It took both the Coral and the Deep Neptune's winches to bring the old looking thing to the surface and haul it back to the docks. The third ship of the fleet. The rising sun returned to port not too long after with her nets only half full and her crew's hopes of the day's catch crushed upon the sight of the mysterious mass sitting in both sister ships' nets, slowly being lifted by the loading dock crane onto a flat bed truck.

Mario:I wonder what in the hell that thing is?

Kenji: I dunno. It looks lie a . . . oh, a giant crescent roll with a bite taken out of the middle.

Mamuro: I know what it looks like jerk; I'm wondering what it IS!

Kenji: Well I can tell you what it isn't . . . a fish! We haven't caught a single fish, since this stupid thing wound up in our nets last week. At this rate, we won't be in business for very long you know. I heard they are hiring at the new mall they built in the city.

Mamuro: You go ahead and take a job doing mall security work, I'm gonna ask about getting some money for this thing. I mean it gotta belong to someone right? Probably a part of that ship that exploded a couple of weeks ago.

Omi: I think it fell from the sky.

Mamuro: From where . . . Mars! Look, it obviously came from that ship that exploded when we picked up your weirdo friend who loves coffee. Whatever country he's from is probably looking for the damned thing and is willing to pay some good coin for it too!

Kenji: That was a Chinese fishing trawler that was struck by this object we're looking at I believe.

Omi: And I believe it too, Kenji . . . and our newly found friend isn't from China.

Kenji: I don't think he is from anywhere.

Mamuro: You two are dreamers . . . you should be trying to figure out how to save your father's company Omi, while it's still here of course.

Omi: And you should be somewhere else working while you still have a job.

Mamuro: We'll get to it.

Kenji: You shouldn't let him get you so angry Omi. He does have some nerve . . . the jerk!

Omi: I'm not angry with him . . . I'm angry because he's right. My dad is using the family savings to pay you guys. Please don't tell anyone . . . He really feels things will pick up and he doesn't want to loose you guys.

Kenji: Even Mamuro?

Omi: Sadly even that dummy serves some purpose here. The only one that has been around longer then him is you.

Kenji: Well there's a reason for that.

Omi: Which is?

Kenji: Loyalty . . . plain and simple and how I respect your father and how he respects the environment. Men like him are hard to come by these days. Everyone like that loser Mamuro is always looking for a new way to get money . . . not caring about his surroundings or the people connected to him . . . it's almost as if him and everyone else on this planet is blind and/or heartless. Well not me Miss Omi. I'll stand by your father and you. I wont tell anyone what you told me. You can trust me.

Omi: Thank you Kenji. I don't know what to say.

Kenji: No need to say anything Miss Omi. Besides, your probably going to stop having to sell fish soon anyway . . . . There won't be any left to sell.

As the days pass, Gaiyymann slowly builds up his strength and acclimates himself to his new alien surroundings. He knows he has to free Ononot from the escape capsule and formulate a plan, but he needs to do this someplace beyond the eyes of humans, from what he has seen so far, he knows that this is the planet they were looking for but the humans are not ready to accept the fact that not only are they not alone in the universe, but their relationship with each other is closer then they could ever imagine.

Akira: He refuses to come home with us. I've tried everything but he wont leave the docks, and he's been acting real odd since we pulled that thing out of the water and stuck it into one of our dry docks.

Omi: He's been eating and drinking like crazy . . .

Akira: Then he mumbles something and then goes into a coma, like he's praying or something.

Omi: Probably hell, I'd be praying too if I was stuck somewhere. With people I couldn't understand. Couldn't you make out what he was mumbling?

Akira: No, he's speaking too low and besides . . . I don't think he wants to talk to us anymore, he hasn't tried after the day we came back from the day we found that thing in the nets.

Omi: He goes to the dry docks sometimes but only by himself. Dad, I don't even think we can help him anymore.

Akira: I agree but we can't give up on hope just yet. If we could just talk to him!

Mamuro: Now listen guys, we all know the old man is not going to be able to keep this business afloat for long and we all have families.

Later on that day, Mamuro spreads the word amongst the crewmembers he can trust to meet behind Bay#3 in groups of 5 as to not bring attention to them.

Seaman#1: Except for you!

Mamuro: Yeah . . . Thanks for the update.

Seaman#2: Let him speak!

Seaman#3: that's right, he's been here longer then all of us.

Seaman#4: If anyone knows what's going on here, Mamuro does!!

Seaman#5: yeah right!

Mamuro: Thanks for the vote of confidence guys. Anyway, we all know that whatever that thing in dry dock is, it's got to be worth something to someone right?

Seaman#2: Yup, that's right.

Seaman#4: It's probably military.

Seaman#3: That's what those ships are looking for!

Seaman#1: If they find out we've been holing it for a whole week we're in big trouble.

Mamuro: Not if we tell 'em we we're forced to by out boss.

Seaman#3: Hey, what about a maritime law, doesn't that thing belong to whomever salvages it . . . I mean no one was found alive on it or anything, we found it . . . . , it should be ours!!

Seaman#5: That only applies to ships bonehead!

Mamuro: And their equipment! It obviously isn't a boat but it might still be some form of equipment.

Seaman#4: The military supersedes all that. Besides, who's going to convince the old man to give it up, and the weirdo who acts like it's his pet gold fish or something? Every time I see that guy he's hugging on it like it has feelings.

Seaman#2: Yeah, weird.

Mamuro: To hell with the both of them! Maybe I can cut a deal with the military but I'll need all of your help to do it.

Seaman#6: The military does not cut deals man. If they want something they come and take it.

Seaman#5: So if it's theirs, why couldn't find it?

Seaman#6: I didn't say it was theirs, I said if they want something . . . .

Mamuro: Never mind that! Stay focused in the money!

Seaman#3: What money?

Mamuro: The money we could get from this thing! Has anyone been paying attention to me?

Seaman#1: I have, sounds like a good plan.

Seaman#6: What plan? The military isn't going to cut any deal with us for all we know they might throw us all in jail just for looking at the thing.

Seaman#2: Don't be paranoid! We don't know what the thing is or how important it is.

Seaman#5: So I guess the army was just on a fishing trip when they stopped and boarded us at gunpoint?

Mamuro: Shut up you two! Look I will try talking to the old man but I'm not promising anything. If he says no then we've got to move to plan B.

Seaman#5: And what the hell is plan B?

Seaman#2: Haven't you been listening? Plan A, have the old man deal. If not when the old man and the weirdo aren't around, we deal with the army guys.

Seaman#6: I told you man the army guys can't deal!!

Mamuro: Will you shut up! I'll deal with both; I just need you guys to be ready to get rid of this thing when the time is right.

Seaman#5: I don't know about going behind the old man's back like that . . . feels like mutiny.

Mamuro: The only reason I called you guys here is because I trust you. Besides, this place ain't going to be around for long and then what are you gonna do, go work for another fishing company? Well I don't know about you but I'm tired of working for people. I've been working for Akira for 15 years now and what has it gotten me . . . a sore back and calloused hands! Well not anymore!! I'm looking out for #1, me!! This is my chance to have something! My father offered to buy me this company years ago to help make it something but Akira wouldn't have it. We'd be right up there with the big guys if the old man weren't so stubborn! It drove my father crazy because he knew they could be a success together! But no . . . my dad used his money and started a rival fishing company that got swallowed up by one of the larger fishing manufactures! They wouldn't even hire us!! My dad asked . . . no, actually begged Akira to hire me because he was too used up by the sea to work anymore!! I'm not ending up like my father, I'll be damned if I do!! Now WHO'S WITH ME!!?

Seaman#6: Hell man, I am!

Seaman#2: Yeah!!

Seaman#3: Ok. I'm in!

Seaman#4: If you think we can do it lets do it!

Saeman#5: I still don't like it. But I like the idea of being a poor ass man even less, I'm in!!

Seaman#1: Ok we're all behind you . . . how do we begin?

Mamuro: Just leave the old man to me and keep an eye on the stranger. We'll need to get rid of him somehow. I know he's a wanted man and all but if I call the cops, we all could get in trouble for harboring a fugitive so we've got to think of another way . . .

Seaman#1: We could always say we didn't know.

Seaman#3: and Omi and the old man could always say we did!

Mamuro: I say we out the stranger back where we found him.

Seaman#5: Now I know I don't like it! You talking about murder!

Seaman#3: What?

Seaman#6: Yeah man, putting him back out there might as well be murder.

Mamuro: I'll give him food and water or something like that . . . a cell phone so he can call, for help, something like that! I'm not going to kill 'em!

Seaman#32: You may not but the sea probably will . . .

Mamuro: Look nobody's going to die ok. Let me, worry about the details you guys just work with me! Trust me, when we're done, we'll all be rich!! None of you will ever have to work again . . . .

Seaman#5: you'd better be right!

That's right, an ominous tone hovered over the home of Dr. Richard Wong. Astro physicist, Dr. Gregory Mitchell, computer systems analyst and Dr. Jonathan Weber, computer program designer, engineer, and chemist/ metallurgist. The four have known each other for years and have had a dream ever since the notion of an object possibly colliding with their dear planet earth, to build a diverse planet capable of removing that threat from the lives of their fellow men, the stellar defense system grid. The world's government believes that the multi billion-dollar defense system failed its first test and allowed something to jeopardize the planet and kill hundreds. They are out to prove otherwise no matter the cost.

Dr. Cocoa: Everyone has their appropriate material?

Dr. Wong: Yes, I have mine.

John: Done.

Gregory: It's right here.

Dr. Cocoa: Good my wife will make copies as we talk. I've mentioned before that I have contacts at the impact zone . . . .

Gregory: That's correct . . .

Dr. Cocoa: Well, I'm heading out this week I've managed to get a clearance from the military. It seems they've got something out there that need a lot of explanations, everyone I know in the science community with a background in astrophysics, meteorology, radiology, quantum physics, biotechnology, chemistry and biology with more then 2 PhD's have been heading to Mongolia these past couple of weeks and none of them have returned, nor have they been allowed to communicate with their loved ones. Something is definitely happening over there that we need to know about.

Dr. Wong: I've been doing some thinking about the defense grid and I've come to a conclusion on my part.

Gregory: Which is?

Dr. Wong: I am sure the same conclusion that you have come up with upon conclusion of your analysis of the defense system.

John: We're running out of time good doctor, please . . . . no games.

Dr. Wong: Believe me gentlemen, no games are being played on my behalf. It seems the only ones here playing games are the joint government administration.

Dr. Cocoa: The J.G.A.? Why is that?

Dr. Wong: Think gentlemen . . . upon reviewing your analysis involving each of the Planetary Defense Grid systems there seems to be one constant, each system seems to be tracking an object that is not traveling a fixed path, even though records clearly state otherwise.

John: I agree. I have confidence in what I do and I stand behind my tracking programs. It couldn't have gone haywire like that.

Gregory: I believe you Johnny boy, my investigations of my weapons programming couldn't have done the same as well. The weapons are not being fired sporadically; if the program had failed, there would be no controlled firing of weapons; in fact there is a 90% possibility that all weapons would have fired simultaneously had there been a program failure not one by one.

Dr. Wong: Which brings me to the defense grid monitoring systems. There seems to be only one discrepancy in all the programs on board the Defense grid and even they, the technicians under me, can't explain why. All the other programs seem to be reacting to an object not on a fixed course, but moving sporadically, almost evasively but my monitoring programs say different. I found that to be odd so I cross-examined them with our old prototype programs and foundation scenarios of possible events pertaining to the defense grid.

Dr. Cocoa: Well what did you find?

Dr. Wong: Nothing absolutely nothing.

John: So why are you talking?

Dr. Wong: Now, now young John . . . , I haven't finished . . .

John: Please gentlemen, don't call me Johnny, my mother calls me that, I hate it . . .

Dr. Cocoa: Johnathan then?

John: No, my father calls me Johnathan . . .

Gregory: Can we please get to the point Dr. Wong!

Dr. Wong: Absolutely my good man . . . .

Dr. Cocoa: My apologies John.

John: Right, right. Thank you Gregory.

Gregory: You're welcome Johnny Boy.

Dr. Wong: Well like I said every program that the previous incident came up against had a negative result, all except one.

John: Which one?

Dr. Wong: The one for the original design for the defense grid.

Gregory: The original wasn't for planetary defense . . .

John: It was for continental defense! I'd completely forgotten, amiss all of this madness!

Dr. Cocoa: That's why it acted the way it did! The system didn't know how to react to a multi-directional moving target because we didn't program that into it!

John: But the initial software design was targeting missiles, high altitude fighters and bombers . . . even defense satellites!

Gregory: Right, so, when the system was introduced to a multi-directional target, it tried to adjust by failing back on its original software!

Dr. Wong: Indeed it did! Just like you wise gentlemen designed it to do.

John: Hot DAMN!! It's not our fault!

Dr. Cocoa: That is what I've been trying to tell you . . .

Dr. Wong: Don't get your hopes up yet, we still don't know why the grid monitoring system tracked a fixed moving object.

Gregory: Honestly Simon, could your monitoring system be flawed?

Dr. Wong: If it is we'll know soon enough. Simulated tests are being run mimicking the Mongolia incident on a mock computer generated defense grid, unbeknownst to our governments at my private facilities in Utah. We should have an answer by the closing of the week.

John: What do you think the test will reveal?

Dr. Wong: Hopefully for us that defense grid was acting with design specs.

Gregory: That also means something else. That what ever that thing was that struck the Mongolian Peninsula wasn't a meteorite.

John: And wasn't ours as in Earth's.

Dr. Cocoa: Thank you dear. Here are our copies of each other's notes, findings, design specs ect. Cross-reference with extreme prejudice gentlemen, if we are going to present this to the J.G.A. we have to be right, evidence must be irrefutable!

Dr. Wong: I have to leave your company gentleman; I have too much to do.

Dr. Cocoa: As do we all. I have to pack. Sweetheart, can you put this all of this on disc for my journey.

John: I guess I'll be leaving too . . . .

Mrs. Cocoa: Now wait one moment gentlemen! This time you are going to stay for a slice if my famous pineapple cake aren't you?

Gregory: you know just what to say Mrs. Cocoa, besides . . . . I could smell it baking before I walked out the door.

John: Only if they are fresh my friend.

Mrs. Cocoa: Well, you'll love it just the same. Kaneohe can you help serve?

Dr. Cocoa: Yes dear . . . I'll be right back.

Gregory: Why all of a sudden the love for pineapple cake?

John: We've got to talk . . .

Gregory: We're talking buddy . . .

John: if this turns out to be some kind of set-up stall tactic . . . .

Gregory: what are you talking about!?

John: SSHH!! Listen, I know someone that got hung out to dry by one of the senior guys in his company.

Gregory: you think Dr. Wong is setting us up?

John: No . . . who has the most to loose here?

Gregory: Who, Kaneohe? Get outta here!. Why? For what reason?

John: For one, it's his baby! He designed the damned thing!

Gregory: I don't believe that, not for one minute!

John: Keep your voice down!

Dr. Cocoa: Believe what?

John: He doesn't believe that I had a violent allergic reaction to pineapples when I was a child.

Mrs. Cocoa: Well I hope you don't feel the need to eat my cake on my account.

John: No, no I'll be fine, I was a child when it happened.

Dr. Cocoa: I'll be right back gentlemen, I need to retrieve a clean disc for my wife before I forget where I placed it again. Enjoy your cake.

Gregory: Thank you doctor, we will.

John: Look, I'm warning you if we don't prepare for the unexpected, then we are both fools!

Gregory: But didn't you believe what Dr. Wong said about the programming sequences and the data?

John: I believe him, but that doesn't change what happened! The J.G.A. is still going to be looking for a scapegoat and little fish that get swallowed by big ones!

Gregory: What!? Please eat your cake!

John: No, I'm serious! Dr. Wong and Dr. Cocoa have known each other for years, we're the new kids on the block. We don't have the connections, money and the history to get out of anything if we get caught up in anything!

Gregory: Listen, I know Dr. Wong, he wouldn't leave us out to dry if something was wrong and besides, we all helped in the construction of this thing, not just us two.

John: I know that, but I also know the bottom guy always gets hit the hardest and what's with Dr. Wong and lawyers!

Gregory: Yeah I know you know a guy.

John: You are not taking me seriously. They both have families, children, money, long standing careers in the science community and we don't. We've got to protect ourselves!

Gregory: Well I don't know about you, but I've got money . . .

John: All right, don't take me seriously! But don't come crying to me when shit hit's the fan and you end up the one-legged man in the ass-kicking contest!

Gregory: you're serious about this aren't you?

John: Hell yeah!

Gregory: All right. Let's pretend that the world was coming to an end and out asses were in a sling. How would these two go about framing us like Mona Lisa?

Dr. Cocoa: Did I miss a good fine art discussion gentlemen?

John: No, I-uh was just telling Gregory about an interesting prank we pulled on one of our dorm mates in senior year. Nothing too funny to repeat though. Look, I really better be going. Thanks for the cake Dr. Cocoa . . .

Dr. Cocoa: Call me Kaneohe John, and I do apologize for the name teasing earlier.

John: Don't worry about it. It was all done in jest. I'll call you later Greg.

Gregory: Please don't call me Greg, my mother calls me that . . . . seriously!

Commander Gault and his assembled team of scientist and soldiers stand at the threshold. No man has ever been down this dark path before. Unlike mans other dark paths, deceit, lies and ill will towards one another. This path is different. This path follows the edge of the complete unknown and descends into the very heart of destiny. This path, once taken, can never be mapped, lit, trusted or back trodden on. This path is a man's future and permanence.

Commander Gault: Dr. Rampashard are you ready?

Dr. Ramp: I am ready, sir.

Commander Gault: Mr. Billings do we have everything all clear on your end?

Mr. Billing: All clear.

Commander Gault: Dr. Liu, are we clear to proceed on your end?

Dr. Liu: Scope is clear on my end. You may proceed when ready.

Commander Gault: Mrs. Gathings are we all clear?

Mrs. Gathings: Systems all clear across the board.

Commander Gault: Dr. Gigot, are you standing by?

Dr, Gigot: I am standing by.

Commander Gault: Dr. Hallstien, are you standing by?

Dr. Hallstien: Standing by sir.

Commander Gault: Dr. Smikle, are you on stand by?

Dr. Smikle: On stand by!

Commander Gault: Dr. Kniles, are you on stand by?

Dr. Kniles: That's correct I am on stand by.

Commander Gault: Sgt. Mayweather . . . . status report.

Sgt. May: All emergency teams are on stand by, rescue team on stand by with equipment checked and prepped. Entry party at staging area are ok to go. Monitoring equipment operational. Surveillance equipment operational drone operational. Back up drone operational, secondary operational systems are green light. You may begin operation sweep one. I repeat you may begin, operation sweep one.

Commander Gault: Lt. Higgins are you and your team ready?

Lt. Higgins: Dr. Dell says he feels like he is going to vomit.

Mr. Gathings: tell him that it's a multi-million dollar suite he's about to barf in.

Commander Gault: Let's concentrate people. We've been preparing for weeks for this one moment; we can't afford to drop the ball. Lt. tell the good doctor he can't back out now, we're already in the history books. You may proceed with your team Lt., any problems and/or you can't continue . . . exit immediately is that understood?

Lt. Higgins: Understood sir.

Commander Gault: God speed Lt.

Lt. Higgins: Thank you commander. O.k. team lets move out, we have two hours of air. Pay attention to the countdown display I the lower left hand corner green visor the numbers will slowly change form the color green to yellow to red as your oxygen depletes. Upon entry we split into two groups of 4. The scout group will stay approximately 30 feet ahead of the recon group and 30 feet behind drone #1. Dr. Dell and I will be in recon group. Sgt. Miller will lead scout groups as planned, understood?

All: Understood sir.

Lt. Higgins: Beginning verbal reports . . . we are entering what we believe is some sort of port way through the starboard slide of the vessel, by the looks of things I'd say the vessel is on a 25 degree decline and tilted approximately 40 to 41 degrees to the starboard side. There are no lights and the needle is not moving in out radiation scanners any more then they did upon crash-landing. I can see scout team ahead . . . . there seems to be a bend up ahead . . . going to the left it seems. The floor seems to be at an angle . . . slight but it meets at the center, it's not flat . . . neither is the ceiling. Drone one seems to have found something, there's a drop off ahead, hold on scout groups. You guys getting the same readings?

Sgt. Miller: Affirmative sir. I think we're in some shaft of vent of some sort.

Commander Gault: Can you confirm that on drone #1 Sgt. Mayweather?

Sgt. May: Getting readings now sir. I can't be to sure sir, my guess is that it is some sort of shaft but we won't know that until we proceed sir.

Sgt. Miller: Lt. I've caught up with the drone. It seems to be some sort of drop off here . . . . I can't tell if we're walking on the floor or the ceiling here . . . there are no points of reference to the fact . . . . we're just coming and going . . . left or right . . . no up or down . . .

Lt. Higgins: Until we come to some sort of instrument panels, we won't be able to tell either, how far is the drop off Sgt . . . I'm coming up on you now.

Sgt. Miller: I can't tell my lights cant seem to penetrate this darkness, officers . . . . Shine your lights in that direction, I want to see something.

Lt. Higgins: Do you see something?

Sgt. Miller: Our lights aren't penetrating this darkness . . . this is crazy! What's the drone reading . . . . I'm getting 0.0 on energy levels here; we should at least be picking up our own suits.

Commander Gault: Well Sgt. Mayweather what's the drone picking up?

Sgt. May: 0.0 sir.

Commander Gault: You heard the man, our end says the same.

Sgt. Miller: WHAT WAS THAT, DID YOU SEE THAT?

Soldier#1: Get back, sir!!

Lt. Higgins: What did you see, did you see anything Dr. Dell?

Dr. Dell: Something moved out there . . .

Commander Gault: STATUS REPORT gentlemen, what's going on in there?

Solider#2: something did move by us . . . floated by the opening sir.

Soldier#3: What in the hell is that!!

Soldier#4: Sir, whatever it is the drone can't pick it up on the sensors.

Soldier#5: LOOK!!

Soldier#6: It's a BODY, there's BODIES floating around in there man, what in the HELL!!

Lt. Higgins: Commander Gault . . . I think we've found a room full of bodies sir, but we can't judge the drop off from the edge of the shaft.

Commander Gault: Sgt. Mayweather, get the back drone on line, Lt. Higgins send in drone 1, if we loose it, drone 2 will be with you 4 minutes after loss is affirmed . . .

Lt. Higgins: Ok boys, here's where it gets interesting, send in drone 1.

Soldier#1: It's floating away! There's some sort of anti-gravity field in the room . . . .

Sgt. May: The drone just went off the scale Commander . . . . we can't get a fix on it. It's gone! No signal at all.

Commander Gault: It's going to switch to secondary protocol Lt. Higgins, have you got the transmitter with you.

Lt. Higgins: Affirmative Commander. We can barley see it . . . . it's still operational.

Dr. Smikle: What's secondary protocol?

Sgt. May: The drone automatically starts to triangulate its surroundings with its own location using all means of sensory perception . . . everything from radar to sonar to spatial tactical uniformly, you know trying to determine corners and such by use of touch.

Soldier#2: That'll take all day, Lt. Higgins, permission to venture inside . . .

Lt. Higgins: That's a negative soldier, we wait for the drone to do its job. We don't go venturing down dark paths without proper precautions being taken.

Dr. Dell: Yes, we're supposed to stay together Doc?

Lt. Higgins: How are you holding together . . . Doc?

Dr. Dell: As well as to be expected Lt.

Sgt. Miller: this is taking to long Lt. For all we know drone 1 is on the other side of the ship right now.

Lt. Higgins: no Sgt., once the transmitter is activated, it'll make its way back here don't you worry.

Soldier#3: How . . . when it can't even get a decent signal here?

Soldier#5: We're dealing with technology we can't begin to understand.

Soldier#1: Not yet anyway.

Lt. Higgins: Cut the chatter . . . . Commander, I'm going to let one of the recon men go in after the drone, with your permission sir.

Commander Gault: We don't want to loose anyone Lt.

Lt. Higgins: That's why he's going to be safety tethered to me sir.

Commander Gault: I can't have you do that to your team leader. Tether him one of the recon team and stand by. Sgt. Mayweather, what is the status on drone 1?

Sgt. May: Still 0.0 on any of its transmissions. It's gone sir.

Dr. Gigot: Can't you strap that soldier to drone 2 Commander?

Commander Gault: Good idea Lt. Higgins, tether the soldier to drone 2 and send him in. 148. Lt. Higgins: Excellent idea commander. You heard the man, let's get to it Sgt. Miller.

Sgt. Miller: Step up Stevens; it's your lucky day soldier.

Soldier#3: I can hardly wait sir.

Soldier#2: Let me go sir.

Soldier#3: Be my guest!

Sgt. Miller: Knock it off!! Ok Brooks, it's on you, snap out of it!

Soldier#2: Yes sir.

Sgt. Miller: I guess we'll have to do this old school . . . . if you have any problems soldier tug the tether and we'll pull you back A.S.A.P. understood?

Soldier#2: Yes sir!

Lt. Higgins: Keep talking as you go in . . . . I need to see what and how we are affected in there. You ready?

Soldier#2: Yes sir!

Lt. Higgins: He's in . . . . what do you see solider?

Soldier#2: I see light all around me . . . . MY GOOD GOD!!

Lt. Higgins: What is it soldier?

Commander Gault: What's the soldiers status ladies and gentlemen?

Dr. Liu: sensors on the suit went dead . . . . I'm not getting anything!

Dr. Gathings: Bio-Rhythmic and brain scan functions have, both flat lined . . . . it's as if someone pulled the plug!

Dr. Ramp: It's obvious what ever goes in that room is its own.

Mr. Billings: He should be alright . . . . those suits were designed to withstand anything we could think of!

Dr. Dell: Pull him back!

Lt. Higgins: Help Me!

Soldier#4: Wait I think I can see him, he's holding something, pull him in . . .

Soldier#5: He's got the drone. He's got the drone with him I can see it!!

Lt. Higgins: I've got him!! . . . . move back, someone secure the drone!

Commander Gault: Give me an update on that soldier people!!

Dr. Gathings: Still flat line . . . . no readings at all . . . he's non responsive, wait a minute his body isn't giving off any body heat.

Mr. Billings: I'm overriding the protective visor.

Lt. Higgins: This is impossible . . . .

Soldier#3: What in the hell is going on around here?

Dr. Dell: We shouldn't go any further!

Commander Gault: Talk to me Higgins!!

Lt. Higgins: His suit is empty.

Commander Gault: What!!?

Lt. Higgins: He's not in his suit, he's gone!

Mr. Billings: That can't be possible, it never opened. You can't get out of that suit in under a minute, it takes at least 30!!

Commander Gault: I'm aborting this mission, Lt. Higgins return to base!

Lt. Higgins: Commander, with all due respect sir, we can't just leave!

Commander Gault: For all we know you could have disintegrated in that suit . . . return to base with your teams Lt. mission is scrubbed until analysis can be made on the drone and the suite. Is that clear!?

Lt. Higgins: Sgt. Miller . . . , take recon and scout team and exit the ship . . . I'm going with drone 2 after Michaels.

Sgt. Miller: Sir we'll all go with you . . . .

Lt. Higgins: This is no time for heroics miller, just do as you were ordered!

Sgt. Miller: But Lt.

Lt. Higgins: No buts Sgt . . . . ,and get yours out of here!!

Dr. Dell: You know if you do find him you won't know what to do with him if he's injured.

Lt. Higgins: You trying to tell me something Dr.?

Commander Gault: Don't you dare go against my orders Lt. you're too valuable to be lost so carelessly!!

Lt. Higgins: I'm sending the teams back with Sgt. Miller, Commander. I'm proceeding into the room sir.

Dr. Dell: well . . . not at first, but I've figured since I've come this far . . .

Lt. Higgins: I was thinking the same thing.

Sgt. Miller: We'll wait for you until our air runs out Lt. You still here Sgt.? Follow protocol and get your men out of danger and that drone data analyzed. That information may come in handy later. Understood?

Sgt. Miller: Understood sir. Alright men lets go. At least take our fire arms Lt.

Lt. Higgins: Why? He didn't take his . . . see still in the suits hand.

Sgt. Miller: Good Luck Lt . . . . , and Dr.

Commander Gault: Lt. when you get back, your ass is in deep shit, you get me!>

Lt. Higgins: Loud and clear Commander. Wish me luck.

Commander Gault: No such thing Lt.

Dr. Dell: Should we hold hands Lt.?

Lt. Higgins: Not a good idea . . .

Commander Gault: If you're not back within 24 hours, it's back to square 1 with this thing, you know that don't you?

Lt. Higgins: I also know we don't have that luxury as well.

Commander Gault: God speed once again Lt.

Lt. Higgins: Thank you Commander.

Dr. Dell: After you Lt . . . . and please, keep me safe in there.

Lt. Higgins: I'll do what I can sir.

Dr. Gigot: I wonder what they will find?

Dr. Gathings: Both Lt. Higgins and Dr. Dell's Suits are showing no life readings Commander Gault.

Mr. Billings: No response from suit sensors Commander.

Dr. Ramp: Continue monitoring procedures Commander?

Commander Gault: Day and Night . . . . I don't want this equipment turned off until they're standing in this room. Is that UNDERSTOOD!

www.ingramcontent.com/pod-product-compliance
Lightning Source LLC
Chambersburg PA
CBHW021903170526
45157CB00005B/1949